I0035450

*First Edition*

# Revising Basic and Clinical Pharmacology

*Dr S. Steele*

$$A M p$$

*Academic Medical Press*

Published by Academic Medical Press, a division of Academic Medical Consulting. Nottingham, UK.

Academic Medical Consulting.
Ebury Road, Carrington, Nottingham NG5 1BB
*Somniare audemus*

Copyright © Dr S. Steele 2012

All rights reserved. No part of this publication may be reproduced or transmitted in any form or by any means, electronically or mechanically, including photocopying, recording or any information storage or retrieval system, without prior permission in writing from the publisher.

This work is registered with the UK Copyright Service: Registration No: 342347

First published 2012

Whilst the advice and information in this book are believed to be true and accurate at the date of going to press, neither the author nor the publisher can accept any legal responsibility or liability for any errors or omissions that may be made.

Any websites referred to in this publication are in the public domain and their addresses are provided by Academic Medical Consulting for information only. Academic Medical Consulting disclaims any responsibility for the content.

ISBN 978-0-9566443-4-3

**Further copies can be obtained from:**          http://www.lulu.com
                                                 http://www.amazon.com

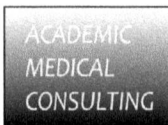

ACADEMIC
MEDICAL
CONSULTING

Preface to the first edition

Pharmacology is unusual amongst the medical disciplines in having an equally strong pre-clinical and clinical presence. However, most medical schools place the bulk of their training in pharmacology *early in their programmes*, leaving students comparatively bereft when they reach the active clinical phase. Accordingly in order to bridge this gap, any text that addresses this subject for medical students must have a real balance of basic *pharmacological science* and pertinent *clinical applications*. This book collects important theoretical learning points in pharmacology and combines them with scenarios that a practicing clinician will recognize, to facilitate a more complete understanding of pharmacological medical practice.

The book is for use by medical students approaching significant pharmacology exams, or by students nearing the end of their medical training. The text is divided into four sessions ("*mock exams*") of mixed *multiple choice* questions, *short answer* questions and *viva voce (oral)* questions. The student is advised to attempt these sessions of questions after they have reviewed their home medical school's course material. These 225 questions have been chosen to ensure a balanced and representative pharmacological view of a typical undergraduate medical course and are intended for revision and active reinforcement of core material.

Finally, I would like to thank my colleagues who reviewed the material in this book for their constructive criticism and their invaluable help in generating this final form and content.

Oxford, September 2012                                    Dr S. Steele

# Contents

If you leave the door open
to success, eventually you
will have the courage to
walk through it.

Sinclair Steele, 2011

# Session 1

# Multiple Choice Questions – Single Best Answer

1) A normally fit and well 75 year old woman presents at A and E "off legs." She is confused and has a temperature of 37.9°C. She has no signs of trauma, no neurological signs, her blood glucose is within normal limits and so are her thyroid function tests. Which of the following are you most likely to prescribe?

A) A cephalosporin
B) 200mg trimethoprim bd po
C) 50mg atenolol od po
D) 10 units of insulin iv
E) 1g paracetamol qds po

2) Which of the following analgesics does not contain codeine?

(A) Co-codamol
(B) Tylenol
(C) Nurofen plus
(D) Anadin extra
(E) Solpadeine max

3) Which of the following is not a side-effect of opiate use?

(A) Constipation
(B) Respiratory depression
(C) Addiction
(D) Mood elevation
(E) Diplopia

4) Which of the following antibiotics does not achieve its therapeutic

effect through the inhibition of cell wall synthesis?

(A) Quinolones
(B) Carbapenems
(C) Vancomycin
(D) Cephalosporins
(E) Penicillins

5) You are called to a ward where an agitated elderly man is physically threatening patients and nurses. The security staff have already been called. The lead nurse asks for help in sedating the patient. Which of the following prescriptions is most helpful in this situation?

A) 5mg haloperidol po
B) 40mg diazepam po
C) 3mg atropine iv
D) 1:1000 adrenaline
E) 2mg haloperidol im

6) A 55 year old woman with a long history of intractable depression is prescribed selegiline to which she responds well. She has no other significant comorbidity or family history. She attends her daughter's wedding, which she plays a major role in organizing. During the wedding she drinks and eats heartily. In the middle of the wedding she complains of chest pain, palpitations, dizziness and headache. Her nose starts to bleed. She becomes drowsy and confused. What is the most likely diagnosis?

A) DVT and PE
B) MI
C) MAOI reaction

D) Malignant hyperthermia

E) Alcoholism

7) In the above scenario the doctor arriving on the scene makes the correct diagnosis and in the acute setting would initially prescribe which of the following?

A) 300mg aspirin po

B) 10mg labetalol iv

C) 40mg nifedipine po

D) Oxygen

E) Sublingual GTN

8) In clinical pharmacological terms, which of the following is the odd one out?

A) Paricalcitol

B) Calcitriol

C) Alfacalcidol

D) Cholecalciferol

E) Cholesterol

9) You have a patient that is apparently thyrotoxic. On examination a goitre is obvious. You submit venous blood to test for T4 and TSH levels. The blood tests show significantly elevated T4 and significantly diminished circulating TSH. What is the most likely diagnosis?

A) Primary hypothyroidism

B) Secondary hypothyroidism

C) Primary hyperthyroidism

D) Secondary hyperthyroidism

E) Secondary hyperparathyroidism

10) The patient described in the previous scenario declines surgery and so a drug treatment is started. Which of the following is most likely to be successful?

A) 0.9mg TSH bd iv
B) 100 intl units calcitonin od sc
C) 10mg salbutamol tds
D) 10mg carbimazole od po
E)  25 micrograms T4 po

11) You are a GP managing a 48 year old diabetic patient. The patient has a total blood cholesterol of 7 mmol/l, a resting blood pressure of 150/95 and an HbA1c > 9%. You elect to start the patient on a statin. Which of the prescriptions below would be most helpful clinically?

A) 40mg pravastin od iv
B) 40mg pravastatin qds po
C) 2.5mg pravastatin od po
D) 20mg simvastatin qds po
E) 40mg simvastatin od po

12) Tamoxifen is

A) An oestrogen receptor agonist.
B) An oestrogen receptor antagonist.
C) A mixed oestrogen receptor agonist and antagonist.
D) A progesterone receptor agonist.
E) A progesterone receptor antagonist.

13) You are a junior doctor called to a medical ward where a 40 year old woman has been suffering from hiccups for the last two days. It is night time and the attending nurse would like you to prescribe something to control the hiccups so that the patient can sleep. Which of the following is most likely to be an effective prescription?

A) 1 litre of normal saline iv
B) 1g paracetamol qds po
C) 20mg buscopan (hyoscine butylbromide) qds po
D) 50mg chlorpromazine tds po
E) 50mg voltarol (diclofenac) tds po

14) Which of the following substances can be used to manage constipation?

A) Lactose
B) Glucose
C) Galactose
D) Lactulose
E) Sucrose

15) A patient with chronic mental health problems has been taking medication for a decade. He has recently developed tardive dyskinesia and oculogyric crises. What class of drug is this patient likely to have been taking?

A) Monoamine oxidase inhibitor
B) Antipsychotic
C) Serotonin specific re-uptake inhibitor
D) Benzodiazepine
E) Opiate

16) Which of the following drugs is correctly matched with its mechanism of action?

A) Atropine; antagonist at nicotinic receptors.
B) Valium; antagonist at GABA$_A$ receptors.
C) Frusemide; osmotic diuretic.
D) Neostigmine; reversible acetylcholinesterase inhibitor.
E) Nifedipine; sodium channel antagonist.

17) Which one of the following drugs has not been used to manage hypertension?

A) Aspirin
B) Frusemide
C) Captopril
D) Atenolol
E) Verapamil

18) Which of the following is a contraindication for the use of streptokinase?

A) Ischaemic stroke within the previous 18 months.
B) Pregnancy
C) Caput medusae
D) Gastritis
E) Renal failure

19) How do thiazides work? By inhibiting

A) The Na$^+$2Cl$^-$K$^+$ transporter on the luminal side of the ascending loop of Henle.

B) The Na⁺/K⁺ pump on the basolateral side of the ascending loop of Henle.

C) The Na⁺Cl⁻ contransporter on the luminal side of the distal tubule.

D) Carbonic anhydrase.

E) Creatine kinase.

20) It is 9pm and you are a junior doctor called to see a 35 year old male inpatient who is claiming to have a severe headache with associated nausea, vomiting and photophobia. He has been an inpatient for three weeks for the management of inflammatory bowel disease. He had a lunch that included oranges and chocolate. The headache has persisted for 7 hours despite his nurse giving him paracetamol. His blood pressure and pulse are normal. On examination no signs of meningism are elicited and no fever is noted. What is the most likely diagnosis?

A) Inflammatory bowel disease
B) Migraine
C) Tension headache
D) Meningitis
E) Encephalitis

21) What drug could you use to treat the patient in the previous question?

A) Paracetamol
B) Labetalol
C) Oestrogen
D) Clozapine
E) Sumatriptan

22) At cardiac arrest situations what dose of atropine is recommended?

A) 1mg
B) 3mg
C) 5mg
D) 10mg
E) 20mg

23) You are called to a ward at 1 am because a patient is having difficulty sleeping. The patient has a history of insomnia with no other relevant comorbidity. What could you prescribe to help the patient sleep?

A) 10mg temazepam od po (nocte)
B) 400mg clozapine od po (nocte)
C) 5mg haloperidol od im (nocte)
D) 1mg risperidone bd po (nocte)
E) 10mg diazepam po od (nocte)

24) As a junior doctor you are called to see a patient who has developed new musculoskeletal back pain, partly because of the extensive period of bed rest. What prescription would you try initially?

A) 400mg ibuprofen tds po
B) 10mg morphine od po
C) 50mg diclofenac od po
D) 1 tablet cocodamol qds po
E) 20mg morphine od po

25) Which of the following is used as a general anaesthetic?

A) Diazepam
B) Halothane

C) Lignocaine
D) Procaine
E) L-dopa

26) Which of the following is not a potential side-effect of (or adverse reaction to) general anaesthetic use?

A) Malignant hyperthermia
B) Myocardial infarction
C) Aspiration
D) Hepatitis
E) Pancreatitis

27) A patient has the following plasma concentrations; $[Na^+] = 120mM$, $[K^+] = 4.7mM$, $[Ca^+] = 2.5mM$ and pH = 7.3. The patient is

A) Hyperkalaemic and acidotic
B) Normokalaemic and alkalotic
C) Hyponatraemic and acidotic
D) Hypercalcaemic and hypokalaemic
E) Hyperkalaemic and alkalotic

28) A patient with severe aortic stenosis is given a mechanical heart valve replacement. Which lifelong drug will the patient require?

A) Prednisolone
B) Benzylpenicillin
C) Warfarin
D) Methotrexate
E) Aspirin

29) After a myocardial infarction a patient is put on regular aspirin doses. What is the prescription most likely to be?

A) 35mg od po
B) 75mg od po
C) 150mg od po
D) 225mg od po
E) 400mg od po

30) A 45 year old patient with HIV complains about a burning sensation in his mouth. On examination raised white plaques are evident. What drug would you use to manage the condition?

A) Cefuroxime
B) Nystatin
C) Penicillin
D) Doxycycline
E) Chloramphenicol

# Short Answer Questions:

## Question 1

A 25 year old woman presents at her GP surgery with a 24 hour history of a very painful left knee. She gives no history of trauma. She has no family history of rheumatoid arthritis. 48 hours ago she was at a wedding in Edinburgh. She has no other medical history or associated symptoms of note.

A) What are your three potential diagnoses (differential diagnosis) for acute severe joint pain in such a person? (3 marks)

B) A fine needle aspirate of the joint is taken and sent to microbiology. As a precaution antibiotic cover is started. If the most likely infective agent is a skin staphylococcus, which family of antibiotics would you prescribe first? (1 mark)

C) Eventually the results come back and indicate that the aspirate was sterile and without neutrophils – there is no evidence of septic arthritis. You consider gout as a cause of the young woman's symptoms. What substance deposited in the joint is directly responsible for causing the joint pain? (1 mark)

D) What would you prescribe as an analgesic initially? What would you prescribe to decrease the chances of future acute episodes? (2 marks)

E) Before prescribing NSAIDS to a patient, what three questions should the patient be asked? (3 marks)

F) After being prescribed an NSAID the patient experienced oliguria and subsequent blood tests indicated renal impairment. The potassium ion concentration rose to 7.5mM. What is the normal range of serum potassium concentrations and what is the clinical significance of 7.5mM? (2 marks)

G) What is the immediate drug management of hyperkalaemia and what is the function of each drug? (8 marks)

## Question 2

A previously well 37 year old motorcyclist is involved in a serious road traffic accident. On examination it is noted that his abdomen is enlarged and shows generalized tenderness. His pulse is 120 bpm and regular - his blood pressure is 80/40mmHg. Abdominal and pelvic X-ray images indicate that he has suffered three fractured ribs and a fractured pelvis.

A) What is the term for his current state and what is the most likely underlying cause? (2 marks)

B) Which two classes of fluid can be used to replace his cardiovascular losses? (2 marks)

C) Give an example of each class. (2 marks)

D) Which class of fluid provides the better prognosis in this and similar situations? (1 mark)

E) As a result of this trauma the patient requires major surgery. Which covering antibiotics would you favour for such surgery? (2 marks)

F) As part of the surgery it is necessary to remove the patient's spleen.

Which drug or class of drug might the patient have to take for the rest of his life? (1 mark)

G) Three days after the surgery the patient complains of left knee pain that is resistant to the post-operative analgesics that he is taking. On examination the left knee is extremely tender, warm, swollen and red. On admission the left knee appeared undamaged by the accident and the patient did not complain of pain. The patient has no history of rheumatoid arthritis and no previous trauma to the knee.
A fine needle aspirate is taken and the fluid sent to microbiology, biochemistry and cytopathology. The presence of uric acid crystals is confirmed.

What is the joint disease that this unfortunate patient is experiencing and what is the likely trigger? (2 marks)

H) Name two drugs that can be used in the acute management of gout and briefly describe their mechanisms of action. (4 marks)

I) Name two classes of drugs that can be used to prevent acute attacks of gout and give an example of each. (4 marks)

# Viva Voce Questions

1) A patient has had a recent clinically confirmed myocardial infarction. The patient is not suitable for angioplasty – what drug treatments are possible?

2) ACE inhibitors are now commonly prescribed medicines. What should an alert doctor be worried about when using these drugs?

3) Name a potassium sparing diuretic.

4) Name two organic nitrates used in the management of cardiac disease.

5) What drug would you use to treat *status epilepticus* acutely?

6) What abbreviation do you use if you want a drug to be given at once?

7) What does the term tocolytic mean?

8) What class of receptor does oxytocin bind to and stimulate physiologically?

9) Which of the following drugs is most likely to help you pass a pharmacology exam; LSD, Ecstasy, Amphetamine, Heroin or Cocaine?

10) Name five classes of diuretics.

11) Name a first line drug used in the treatment of cystitis.

12) You are on an ophthalmology rotation and you want to dilate a patient's pupil to examine the retina fully. Classically which drug is generally used?

13) Do glucorticoids preferentially decrease the frequency or intensity of asthma attacks?

# Session 2

# Multiple Choice Questions – Single Best Answer

31) A 20 year old female patient is an emergency admission to hospital after having taken 2 bottles of paracetamol and written a suicide note; the events happening within the last 6 hours. Which organ is most susceptible to acute toxicity?

A) Kidney
B) Heart
C) Pancreas
D) Spleen
E) Liver

32) Considering the scenario in the previous question, what drug would you wish to prescribe to counteract the paracetamol overdose?

A) Omeprazole
B) Cimetidine
C) Naloxone
D) N-acetylcysteine
E) Lansoprazole

33) An emaciated 50 year old man was found collapsed in his flat on a Saturday evening. No one had seen him since the previous Friday. He has a social history that includes unprotected sex with multiple partners and recreational drug use. On examination of the patient you notice needle marks on his forearms and feet. The patient's breathing is slow and shallow and his pulse is 50 bpm. His pupils have contracted to pin points bilaterally (miosis). Which of the drugs below would be most likely to result in a positive clinical outcome?

A) Naloxone
B) Codeine
C) Tramadol
D) Fentanyl
E) N-acetylcysteine

34) Which of the following is not a compound that is used to treat iron deficiency anaemia?

A) Iron fumarate
B) Iron citrate
C) Iron sulphate
D) Iron gluconate
E) Iron nitrate

35) A patient with epigastric pain is investigated by endoscopy and biopsies are taken. On microscopic examination the morphological features of a Helicobacter Pylori infection are noted. You elect to treat the patient with two different antibiotics and a proton pump inhibitor, omeprazole. What dose of omeprazole would you prescribe?

A) 2.5mg od po
B) 5mg od po
C) 10mg od po
D) 20mg od po
E) 40mg od po

36) Which one of the following abbreviations means *four times daily*?

A) qh
B) qds

C) tds

D) nocte

E) od

37) You are called to the bedside of 76 year old female patient who is a longstanding patient on a rehabilitation ward. The attending nurse is concerned by the patient's recent blood results; $[K^+]$ = 7.0mM. Nothing in the patient's history accounts for this result. You elect to treat the patient by giving (i) 10ml of 10% calcium gluconate iv (ii) 50ml 50% glucose iv with 15 units of insulin iv (iii) 30g calcium resonium per rectum (pr). Which of the following most accurately describes the function of calcium gluconate in this scenario?

(A) It stabilizes the electrical excitability of the heart.

(B) It stabilizes the electrical excitability of smooth muscle.

(C) It stabilizes the electrical excitability of nerve tissue.

(D) It stabilizes the electrical excitability of skeletal muscle.

(E) It has a placebo effect.

38) In the above scenario what effect does giving insulin have on plasma $[K^+]$?

(A) It raises the plasma potassium concentration.

(B) It lowers the plasma potassium concentration.

(C) It has no effect on the plasma potassium concentration.

(D) It causes haemolysis of red blood cells to raise the plasma potassium concentration.

(E) It transiently raises the plasma potassium concentration and then returns it to physiological normality.

39) In the above scenario what is the function of the calcium resonium?

(A) It binds calcium ions and exchanges them with chloride ions.
(B) It binds calcium ions and exchanges them with potassium ions.
(C) It binds potassium ions and allows them to be absorbed by erythrocytes.
(D) It binds potassium ions and allows them to be expelled from the body.
(E) It binds calcium ions and exchanges them with sodium ions.

40) As a result of your intervention in the above scenario, the patient survives. On the next ward round your consultant congratulates you. Later that day her registrar asks you to check the patient's medication to ensure that nothing that is being taken can raise potassium levels. Which of the following drugs should you stop?

(A) Prochlorperazine
(B) Methotrexate
(C) Prednisolone
(D) Amiloride
(E) Paracetamol

41) Which of the following drugs has a well established acute side-effect of producing arrhythmias?

(A) Pravastatin
(B) Paracetamol
(C) Lignocaine
(D) Gold
(E) Penicillin

42) Co-trimoxazole is used to treat:

(A) Pneumocystis jiroveci infections
(B) Hayfever
(C) Gout
(D) Oral candida albicans infections
(E) Cushing's disease

43) Pyridostigmine is useful in the treatment of myasthenia gravis because it:

(A) Attacks the antibodies that cause the disease.
(B) It binds to acetylcholine.
(C) It causes acetylcholinesterase to be endocytosed.
(D) It inhibits acetylcholinesterase.
(E) It stimulates the sympathetic nervous system.

44) You are called to the bedside of a patient with a history of episodic atrial fibrillation. The patient's pulse is 155bpm. The patient's level of consciousness is currently reduced and has been varying for the last 48 hours in concert with the episodes of atrial fibrillation. You decide to use intravenous amiodarone to attempt to return the patient to sinus rhythm ("chemical cardioversion"). What are the clinically important molecular targets of amiodarone?

(A) Bicarbonate channels
(B) Chloride channels
(C) Potassium channels
(D) Sodium channels
(E) Calcium channels

45) A patient is suspected to have taken an overdose of warfarin and laboratory results confirm a very high INR. Which of the following would be least helpful in reversing the effects of the warfarin?

(A) Vitamin K
(B) Fresh frozen plasma and intravenous vitamin K
(C) Fresh frozen plasma
(D) Prothrombin complex concentrate and aspirin
(E) Prothrombin complex concentrate

46) Whilst you are examining a teenager with appendicitis, the patient in the adjacent bed starts fitting. The attending nurse tells you that the patient was admitted urgently two days earlier because of a head injury and is awaiting the results of a CT scan. You elect to deal with the pressing issue of the fit by giving which of the following drugs?

(A) Morphine
(B) Droperidol
(C) Haloperidol
(D) Diazepam
(E) Midazolam

47) You have a cousin who has always suffered from excessive sweating and seeks your advice in managing the problem. He has tried normal antiperspirants but they were ineffective. Which of the following is most likely to be helpful?

(A) Aluminium chloride
(B) Botulinum toxin
(C) Wearing fewer clothes
(D) Noradrenaline

(E) Neostigmine

48) A patient who has just started chemotherapy is feeling nauseous. She asks for something to prevent her from feeling "sick" or vomiting. Which of the following is most likely to be helpful?

A) Domperidone 12.5mg im
B) Cyclizine 50mg tds iv/im or po
C) Aluminium hydroxide
D) Haloperidol 3mg im
E) Drinking a lot of water.

49) Which of the following pairings of drugs and mechanisms of action, is incorrect?

(A) Lactulose – osmotically active laxative that increases water retention in the stool.
(B) Digoxin – antiarrhythmic that inhibits the $Na^+/K^+$-ATPase in the cell membrane.
(C) Chloramphenicol - binds to the 50S ribosomal subunit and inhibits the peptidyl-transferase.
(D) Metformin – increases insulin synthesis in the β cells of the Islets of Langerhans.
(E) Spironolactone – blocks the mineralocorticoid receptor, which would otherwise stimulate sodium re-uptake and potassium loss in the collecting duct.

50) A 36 year old intravenous drug user spikes a fever of 38.5°C and on the same day develops a new murmur. He has his own native heart valves (not synthetic or transplanted valves). You decide that this is an acute presentation of bacterial endocarditis and send off blood cultures.

Before the cultures return you initiate treatment with flucloxacillin (2g 4-hourly iv) and gentamicin (80mg 8-hourly iv). How does the gentamicin work?

(A) It binds to mRNA to prevent protein synthesis.
(B) It binds to the 50S ribosomal subunit in the affected bacteria and causes misreading of mRNA and aberrant initiation of protein synthesis.
(C) It binds to the 30S ribosomal subunit in the affected bacteria and causes misreading of mRNA and aberrant initiation of protein synthesis.
(D) It binds to tRNA to prevent protein synthesis.
(E) It inhibits transcription in the affected bacteria.

51) You are called to see a patient on long term chlorpromazine. You are concerned that the patient may be suffering from the side-effects of this drug. Which of the following is an established side-effect of chlorpromazine?

(A) Bradycardia
(B) Diarrhoea
(C) Mania
(D) Urinary incontinence
(E) Tardive dyskinesia

52) To which class of drug could hydrocortisone be most accurately assigned?

(A) Toxin
(B) Glucocorticoid
(C) Mineralocorticoid

(D) Corticosteroid

(E) Hormone

53) You are called to the bedside of a patient that you notice has a round puffy face, abdominal striae and a buffalo hump. Which drug is he likely to have been taking for a significant period of time?

(A) Prednisolone

(B) Noradrenaline

(C) L-dopa

(D) Growth hormone

(E) Actrapid

54) What is the best description of cocaine's mechanism of action?

(A) Potentiates dopaminergic, noradrenergic and serotoninergic transmission in the CNS by inhibiting their re-uptake from the synaptic spaces.

(B) Inhibits dopaminergic, noradrenergic and serotoninergic transmission in the CNS by stimulating their enzymic degradation.

(C) Potentiates GABA transmission in the CNS.

(D) Inhibits cholinergic transmission in the CNS.

(E) Potentiates L-dopa transmission in the basal ganglia.

55) Which of the following is neither a natural nor a synthetic catecholamine?

(A) Isoproterenol

(B) Noradrenaline

(C) Adrenaline

(D) Dopamine

(E) Acetylcholine

56) A patient is required to take digoxin regularly to manage persistent atrial fibrillation with a fast ventricular rate. Although the drug is being used successfully the patient wants to know why so many blood tests are required. Your answer should refer to:

(A) The narrow therapeutic window of digoxin.
(B) Monitoring for an allergic response.
(C) Evidence of rhabdomyolysis.
(D) Measuring albumin levels.
(E) Measuring free $Fe^{2+}$ levels.

57) Which of the following pairings of drugs and mechanisms of action, is incorrect?

(A) Bupivacaine; inhibits voltage sensitive sodium channels.
(B) Suxamethonium; causes muscle paralysis by acting as an agonist at nicotinic receptors in the neuromuscular junction.
(C) Ecstasy (MDMA); the euphoria is caused by inhibition of serotonin re-uptake in the CNS and increased serotonin levels in the synaptic cleft.
(D) Diclofenac; a specific COX-1 inhibitor.
(E) Pravastatin; inhibits cholesterol synthesis and decreases circulating LDL levels.

58) An adult patient has cystic fibrosis and takes enzyme supplements to maintain near normal digestion. Which of the following enzymes would you most expect to be in the supplement?

(A) Amylase

(B) Pepsin
(C) Lipase
(D) Protein kinase A
(E) Protein kinase C

59) An overweight 50 year old woman has recently undergone a total abdominal hysterectomy with bilateral salpingo-oophorectomy (TAH and BSO). Three days after the procedure she complains about pain and swelling in her left calf. You confirm the localized swelling and a D-dimer test is positive. Compression stockings are applied to the left leg. You start her on heparin and warfarin anticoagulation. What is the mechanism of action of heparin?

(A) Heparin binds to antithrombin III and increases its activity.
(B) Heparin binds to antithrombin III and decreases its activity.
(C) Heparin prevents the synthesis of vitamin K dependent clotting factors.
(D) Heparin increases the synthesis of vitamin K dependent clotting factors.
(E) Heparin hydrolyses fibrin.

60) Why is heparin only given as an initial treatment?

(A) Because it is faster acting than warfarin and is usually given as an injection or as a drip whilst the patient is in hospital.
(B) Because heparin is cheaper than warfarin.
(C) Because heparin has a low bioavailability.
(D) Because heparin is hepatoxic after three weeks.
(E) Because heparin is cardiotoxic after two weeks.

# Short Answer Questions:

## Question 3

A) As a junior doctor working for a surgical firm, you have to manage many postoperative patients. It is not unusual for one of your patients to be pyrexial.
Name four causes (diagnoses) of postoperative fever. (4 marks)

B) What is the first line drug used in the treatment of a urinary tract infection (UTI)? (1 mark)

C) What are the first line drugs used in the treatment of deep venous thrombosis (DVT)? (2 marks)

D) Name the tests used to show that heparin is within its therapeutic range and that warfarin is within its therapeutic range. (2 marks)

E) A 55 year old patient has undergone a heart valve replacement with an artificial valve. As a result she is required to maintain lifelong anticoagulation with warfarin. The aetiology of her heart valve dysfunction is believed to be childhood rheumatic fever. For the first year after the operation the warfarin mediated control of INR is routine. However, after the second year progressively larger doses of warfarin are required to maintain the target INR. Interestingly, after the second year the patient's renal function and cardiac function remain good, however the hepatic function deteriorates insidiously. The patient reports that she was divorced by her husband shortly after the valve transplant and is currently dating again. When she puts down her handbag the physician hears the clinking of glass bottles.

What is the most likely explanation for deterioration in the control of her INR? (3 marks)

F) If the warfarin levels are maintained at too high a level, to what types of complications will the patient be prone? (1 mark)

G) Over the course of the next five years, she develops spider naevi, caput medusae and haemorrhoids.  Blood tests show markedly raised AST, ALT and ALP. Concurrently her personal and social life leave her unhappy and she admits that drinking excessive alcohol is a problem. What affective disorder can chronic alcohol abuse cause? (1 mark)

H) The GP is concerned about the progression of her depression and decides that she needs support from antidepressants.  He considers using an SSRI, SNRI and MAOI.
Give an example of each of these classes and briefly describe the primary mechanism of action. (6 marks)

## Question 4

A shot and wounded soldier is taken to an emergency medical facility. Because of her significant blood loss, it is decided to give her a transfusion. The soldier is unconscious and is unable to state her blood type, nor is she carrying anything that mentions her blood type.

A) She needs at least three units of blood. What blood type would you give her? (2 marks)

B) In the makeshift hospital there is a chart that matches recipients to acceptable donors. However, not all of the boxes in the chart can be read. Complete the empty boxes:
(8 marks)

| Red blood cell compatibility table | | | | | | | | |
|---|---|---|---|---|---|---|---|---|
| Recipient | Donor | | | | | | | |
| | O- | O+ | A- | A+ | B- | B+ | AB- | AB+ |
| O- | ✓ | | | ✗ | ✗ | ✗ | ✗ | ✗ |
| O+ | ✓ | | ✗ | ✗ | ✗ | ✗ | ✗ | ✗ |
| A- | ✓ | | ✓ | ✗ | ✗ | ✗ | ✗ | ✗ |
| A+ | ✓ | | ✓ | ✓ | ✗ | ✗ | ✗ | ✗ |
| B- | ✓ | | | | | | | |
| B+ | ✓ | | ✗ | ✗ | ✓ | ✓ | ✗ | ✗ |
| AB- | ✓ | | ✓ | ✗ | ✓ | ✗ | ✓ | ✗ |
| AB+ | ✓ | | ✓ | ✓ | ✓ | ✓ | ✓ | ✓ |

C) Shortly after the patient is transfused she spikes a fever of 38.5°C and becomes tachycardic, tachypnoeic and hypotensive. You decide that the patient is suffering from an acute transfusion reaction.

How is this most likely to have occurred? (3 marks)

D) The transfusion is stopped, an iv saline infusion is commenced and furosemide given. An expert on blood transfusions is contacted. The subsequent disseminated intravascular coagulation (DIC) is managed by transfusing platelets. Complete the donor/recipient transfusion table below to minimize acute and chronic complications: (7 marks)

| Recipient | Donor | | | |
|-----------|-------|-----|-----|-----|
|           | O     | A   | B   | AB  |
| O         | ✓     | ✗   | ✗   | ✗   |
| A         | ✗     |     | ✗   | ✗   |
| B         |       |     | ✓   |     |
| AB        | ✗     |     |     |     |

# Viva Voce Questions

14) In theory, what types of effect can a failing liver have on drug action?

15) Before a gastroscopy an anxious patient is given midazolam. What type of drug is midazolam and how does it act?

16) For which disease is the drug cocktail HAART used?

17) Define drug efficacy.

18) Who is at most risk of diabetic ketoacidosis?

19) Name a class of drug that will never start fires.  ;)

20) Name four classes of drugs that can be used to manage hypertension.

21) Methotrexate is an anticancer drug which is an anti-metabolite. What is an antimetabolite?

22) Methotrexate can have a notorious adverse effect on the lungs. What is this effect?

23) Why do loop diuretics predispose to hypokalaemia?

24) Name two classic uses of loop diuretics.

25) Which important sympathetic function is controlled by postganglionic muscarinic receptors?

26) Which drug is longer acting, salmeterol or salbutamol?

# Session 3

# Multiple Choice Questions – Single Best Answer

61) Intrathecal drug delivery implies the introduction of a needle into the subarachnoid space for the purpose of instilling a material for diffusion throughout the spinal fluid (CSF). Which of the following is a recognized clinical use of intrathecal drug delivery?

(A) Thrombolysis of CNS thrombi.
(B) Treating encephalitis.
(C) Treating meningitis.
(D) Control of chronic pain, both cancer and non-cancer related.
(E) Managing severe migraines.

62) What is the mechanism of action of heroin?

(A) Heroin acts indirectly at μ opiate receptors in the CNS to remove GABAnergic inhibition of noradrenergic neurotransmission.
(B) Heroin acts directly at μ opiate receptors in the CNS to remove GABAnergic inhibition of noradrenergic neurotransmission.
(C) Heroin acts directly at μ opiate receptors in the CNS to remove GABAnergic inhibition of dopaminergic neurotransmission.
(D) Heroin acts indirectly at μ opiate receptors in the CNS to remove GABAnergic inhibition of dopaminergic neurotransmission.
(E) Heroin potentiates serotoninergic neurotransmission.

63) During a ward round, one of your asthmatic patients has a severe asthma attack. You initiate the protocol for managing such an attack and give (a) 60-100% oxygen, (b) 5mg salbutamol plus 500μg ipratropium, (c) 50mg prednisolone po and 100mg hydrocortisone iv. The patient recovers from the asthma attack. What is the purpose and mechanism of action of ipratropium?

(A) It is an anti-inflammatory that decreases the frequency of severe asthma attacks.

(B) It is a muscarinic that causes bronchodilation and prevents secretion from the bronchial epithelia.

(C) It is an antimuscarinic that causes bronchodilation and prevents secretion from the bronchial epithelia.

(D) It is a sympathomimetic that causes bronchodilation and prevents secretion from the bronchial epithelia.

(E) It is a β-adrenergic antagonist that causes bronchodilation and prevents secretion from the bronchial epithelia.

64) Considering the above scenario, which adrenergic receptors are stimulated by salbutamol to cause the clinically useful effect?

(A) $\alpha_1$

(B) $\alpha_2$

(C) $\beta_1$

(D) $\beta_2$

(E) $\beta_3$

65) If, in the scenario above, the patient had not fully recovered after 15-30 minutes then 250mg of intravenous aminophylline could have been administered. What is the function and mechanism of action of aminophylline?

(A) It is an anti-inflammatory that inhibits COX-1.

(B) It is a bronchodilator that acts through inhibition of phosphodiesterases.

(C) It is an anti-inflammatory that decreases bronchial secretions.

(D) It is a bronchodilator that acts through the stimulation of adenylate cyclase.

(E) It is a bronchodilator that acts through the stimulation of guanylate cyclase.

66) A seven year old boy sees his GP several times in a year because of minor infections. He becomes noticeably more fatigued. On examination his liver and spleen are mildly enlarged and generalized lymphadenopathy is noted. Blood tests show anaemia. The GP is suspicious of acute lymphoblastic leukaemia (ALL). Blood smears and cytogenetics confirm the GP's suspicions. After referral to an oncologist the boy is started on a remission chemotherapeutic regime of vincristine, prednisolone, L-asparaginase and daunorubicin. What is the most likely complication of vincristine use?

(A) Diarrhoea
(B) Hypernatraemia
(C) Hair loss
(D) Blindness
(E) Central neuropathy

67) Why are four drugs used in the remission phase of this chemotherapy?

(A) Because four drugs are used in the maintenance phase of treatment.
(B) To increase the chances of rapidly killing at least 95% of the malignant cells.
(C) Using five drugs would be prohibitively expensive.
(D) To increase the chances of the development of drug resistant strains of the mycobacteria.
(E) To prepare for CNS prophylaxis by radiation therapy.

68) What is the mechanism of action of L-asparaginase?

(A) It increases the basal metabolic rate of the tumour cells.
(B) It facilitates the synthesis of the toxic substance asparagine.
(C) It potentiates the action of corticosteroids.
(D) It potentiates the action of daunorubicin.
(E) It hydrolyses asparagine that the leukaemic cells need for protein synthesis and growth and so is cytotoxic.

69) Which of the following drugs is not used in the initial treatment of TB infections?

(A) Erythromycin
(B) Isoniazid
(C) Ethambutol
(D) Pyrazinamide
(E) Rifampicin

70) Pethidine is a member of which of the following classes of drug?

(A) Anti-emetic
(B) Antidepressant
(C) Opioid analgesic
(D) NSAID analgesic
(E) Sedative

71) A 45 year old man presents with severe dehydration and fatigue at your Accident and Emergency department. When you measure his blood glucose it is 35mM, there is a significant increase in ketones in his blood and the pH of the arterial blood is 7.25. You diagnose diabetic ketoacidosis (DKA) and initiate rehydration, potassium supplementation and start intravenous insulin administration. Why is it necessary to give insulin?

(A) Classically DKA occurs in type 1 diabetics; these individuals cannot produce significant endogenous insulin.
(B) Individuals with DKA produce excessive amounts of proteases that degrade the insulin protein.
(C) Insulin is given to control the plasma potassium levels.
(D) Insulin is given to control the plasma calcium levels.
(E) Insulin is given to control the plasma sodium levels.

72) You have admitted an adult patient with an erratic lifestyle who indulges in intravenous drug use and unprotected sex. On examination, anogenital ulcers are noted. Blood tests reveal treponeme-specific antibodies. You should initiate treatment with which of the following prescriptions?

(A) Doxycyline 200mg bd po for 28 days
(B) Cefuroxime 250mg bd po for 7 days.
(C) Co-amoxiclav 1.2g tds iv for 10 days.
(D) Metronidazole 2g od po for 10 days.
(E) Erythromycin 250mg qds po for 7 days.

73) A patient is on an oncology ward because of disseminated malignant melanoma. Overnight the patient complains of a new severe headache and associated nausea and vomiting. Ophthalmoscopy reveals enlarged retinal veins, loss of venous pulsation and haemorrhaging next to the optic disk. You give 150ml of 20% mannitol intravenously over 10 minutes. You increase the patient's diamorphine dosage and add 8mg dexamethasone at 12 hourly intervals, to be administered intravenously. Finally, you request a surgical consultation. What is the purpose of the mannitol?

(A) It is an osmotic diuretic that decreases extracellular fluid volume so lowering intracranial pressure.

(B) It is a loop diuretic that decreases extracellular fluid volume so lowering intracranial pressure.

(C) Mannitol is an effective analgesic with a short halflife.

(D) Mannitol is an anti-inflammatory medication.

(E) Mannitol is a cytotoxic chemotherapeutic agent.

74) What was the purpose of the dexamethasone in the previous scenario?

(A) It is a glucocorticoid that decreases the inflammation and hence inflammatory oedema, caused by the intracranial malignancy.

(B) It is a COX-1 inhibitor and an anti-inflammatory.

(C) It is an osmotic diuretic.

(D) It is a mineralocorticoid that decreases the inflammation and hence inflammatory oedema, caused by the intracranial malignancy.

(E) It is a radiotherapeutic agent.

75) What is another name for diamorphine?

(A) Heroin

(B) LSD

(C) Amphetamine

(D) Methadone

(E) Ecstasy

76) A 16 year old boy who is a keen footballer presents to Accident and Emergency with a unilateral rash over the left shin. The rash is well defined, erythematous (red), painful and warm. There is no other medical history of note. You make the diagnosis of cellulitis. Which is the most appropriate drug treatment?

(A) Flucloxacillin and benzylpenicillin
(B) Ceftazidime
(C) Ceftazidime plus metronidazole
(D) Trimethoprim
(E) Benzylpenicillin and metronidazole

77) Pilocarpine is a drug used to treat glaucoma. It acts through the stimulation of muscarinic receptors. Based on this information which of the following side-effects do you think pilocarpine would cause?

(A) Bronchodilation
(B) Pupillary dilation
(C) Decreased sweating
(D) Bradycardia
(E) Decreased salivation

78) Which of the following is not a drug used to manage Parkinsonism?

(A) Bromocriptine
(B) Amantadine
(C) Selegiline
(D) Domperidone
(E) L-dopa

79) A 65 year old man has a heart valve replacement that requires him to maintain appropriate anticoagulation levels, using warfarin as an anticoagulant. Over the last year his original warfarin dose has become less effective and this patient's INR (measure of anticoagulation) has begun to fall. You believe that the patient has been compliant with the medication and is not taking any recreational drugs. There has been no significant change in his lifestyle – he continues to be an outgoing man

with a lot of friends. His favourite activities are fishing and visiting the local pub. You decide to take a panel of baseline blood tests for renal and hepatic function. The tests show abnormal function of a major organ. What is the most likely cause of the decreased effectiveness of the warfarin?

(A) Chronic alcohol abuse has stimulated P450 in the liver to increase the clearance of warfarin.
(B) Chronic alcohol abuse has stimulated P450 in the liver to decrease the clearance of warfarin.
(C) Cirrhosis has caused an increase in the first pass effect.
(D) Impaired renal function has increased the warfarin excretion.
(E) Impaired liver function has decreased the warfarin excretion.

80) Phase I drug reactions occur in the

(A) Heart
(B) Kidney
(C) Bladder
(D) Liver
(E) Brain

81) "The branch of pharmacology concerned with the movement of drugs within the body" is the definition of:

(A) Pharmacodynamics
(B) Pharmacokinetics
(C) Drug-receptor theory
(D) Pharmacosolution
(E) Pharmacodiffusion

82) In pharmacology affinity refers to:

(A) The capacity of the drug to inactivate the receptor.
(B) The strength of the bond/attraction between the antagonist and receptor.
(C) The strength of the bond/attraction between the drug and receptor.
(D) The capacity of the drug to activate the receptor.
(E) The inverse of the speed of drug elimination.

83) Which of the following does not refer to drug receptor interactions?

(A) Compliance
(B) Agonist
(C) Affinity
(D) Antagonist
(E) Cooperativity

84) Which of the following is an SSRI?

(A) Senna
(B) Lithium
(C) Venlafaxine
(D) Selegiline
(E) Fluoxetine

85) Which is the most accurate primary mechanism of action of the combined oral contraceptive pill?

(A) Prevents implantation of the embryo.
(B) Suppresses release of gonadotrophins and ovulation.
(C) Prevents fertilization of the ovum by the sperm.

(D) Causes dyspareunia.

(E) Increases the frequency of miscarriages.

86) Which of the following is not a side-effect of HRT therapy?

(A) Increased risk of thrombotic disease.

(B) Increased risk of uterine malignancy.

(C) Increased risk of a pulmonary embolus.

(D) Increased risk of a stroke.

(E) Increased risk of bladder malignancy.

87) The proprietary name (brand name) for salbutamol is

(A) Rentopin

(B) Varent

(C) Salbutamone

(D) Ventolin

(E) Ergatol

88) Which disease/disorder is incorrectly matched with its drug management?

(A) Rheumatoid arthritis - Celecoxib

(B) Myocardial infarction - Aspirin

(C) Pulmonary embolus - Celecoxib

(D) Osteoarthritis - Paracetamol

(E) Febrile episode – Paracetamol

89) Alcohol has which of the following effects when used *acutely*?

(A) Inhibits ADH release.

(B) Causes cirrhosis of the liver.

(C) Causes alcoholic cardiomyopathy.

(D) Causes haematemesis.

(E) Causes haemorrhage.

90) You have a patient who has been smoking for 10 years and is finding it very difficult to stop. She asks you to explain the mechanism of action of cigarette smoke and why it is so difficult to quit. Which of the following statements is most important to include in your explanation?

(A) Nicotine from cigarettes makes breathing easier.

(B) Nicotine binds to nicotinic receptors in the brain to increase the levels of the neurotransmitter GABA in the pleasure centre of the brain.

(C) Nicotine can raise the metabolic rate and cause weight loss.

(D) Nicotine increases the level of the neurotransmitter dopamine in the pleasure centre of the brain.

(E) Nicotine can cross the blood brain barrier.

# Short Answer Questions:

## Question 5

A 58 year old businessman presents at Accident and Emergency complaining of upper abdominal pain. The pain has been present for three months and has worsened over the last week. He waited until after his merger before seeing a doctor; he has been taking Alka-Seltzer but this is becoming less effective. The pain was dull and is now sharp, is 7/10 in intensity and is associated with a disproportionate sense of fullness after a small meal. There is also associated nausea without vomiting. The businessman suffers from osteoarthritis of his left knee and regularly takes ibuprofen.

On examination epigastric tenderness is found. No other abnormalities are revealed. The patient is apyrexial with a regular pulse of 80 bpm and a blood pressure of 135/89.

A) What is the active component of the medication the businessman used to manage his stomach upset? (1 mark)

B) Considering *this* scenario, name three likely causes of the most probable gastric diagnosis. (3 marks)

C) The patient is sent to endoscopy for examination of the stomach and lower oesophagus. Before the investigation, he is given midazolam. Where and how does this drug work? (3 marks)

D) The person administering the midazolam makes a dilution error and accidentally administers 100x the normal dose to this adult patient.

Name four symptoms of midazolam overdose.  (4 marks)

E) The medical team move rapidly to counteract the overdose by giving the patient intravenous flumazenil. Flumazenil acts by competitively inhibiting the binding of benzodiazepines at the GABA$_A$ receptor. The patient recovers and is well.

The endoscopic biopsies that were taken are sent to histopathology and the report is eventually retrieved. It indicates that there is a malignancy underlying the gastritis. A surgical referral is made and the tumour excised and the patient is started on chemotherapy.

Name the two important characteristics of an idealized cocktail of chemotherapeutic drugs. (2 marks)

F) There are four subgroups of classical anticancer drugs that are distinguished by their mechanisms of action. Name the four subgroups of classical anticancer drugs. (4)

G) The chemotherapeutic regime includes herceptin, 5-fluorouracil and doxorubicin. Briefly describe the mechanism of action of each drug. (3 marks)

## Question 6

A 47 year old amateur sportsman is admitted to his local Accident and Emergency department after complaining of chest pain. It is Sunday night and he spent the afternoon playing squash. The previous day he played football for three hours. The chest pain is described as 7/10 in intensity, central and radiates to his neck. It started two hours after he finished his game of squash, and is unusual in not being relieved by ibuprofen. The pain has continued and is still present on admission.

He has no family or personal history of arrhythmia, no evidence of hypertrophic obstructive cardiomyopathy (HOCUM) and no evidence of hyperlipidaemia. He has never smoked. He has no personal or family history of diabetes mellitus. His parents are in their 70s and are in good health – neither has suffered from angina or a myocardial infarction.

Palpation of the ribs or sternum could not elicit the pain.

The junior doctor managing the patient decides that the most likely diagnosis is a myocardial infarction.

A) Name four drugs routinely used to manage a myocardial infarction. (4 marks)

B) Clopidogrel and enoxaparin share a major side-effect (adverse effect). What is it? (1 mark)

C) Describe the mechanism of action of enoxaparin. (3 marks)

The junior doctor requests an ECG and takes blood to test for the release of cardiac enzymes/troponin. The ECG changes are indeterminate with no clear ST elevation. However, troponin and CK-MB levels are significantly raised in the venous blood.

D) Why would a myocardial infarction raise circulating levels of troponin and CK-MB? (1 mark)

E) Although the patient finds the drug regime effective, he complains that one of the medications gives him a headache. Which medicine is it likely to be and why is the headache occurring? (2 marks)

F) How and why are streptokinase and tissue plasminogen activator useful in the management of a myocardial infarction? (4 marks)

G) Three days after admission the patient is still suffering from a similar pain and has developed a pyrexia of 38°C. Angiography indicates that the three major coronary arteries are each more than 90% patent. The relatively young age and paucity of risk factors caused the consultant supervising the junior doctor to reconsider the working diagnosis. What is your new favoured diagnosis? (2 marks)

H) What medications would you use to manage the latest diagnosis? (3 marks)

# Viva Voce Questions

27) Name three glucocorticoid drugs.

28) Can you name the class of drugs to which all of the following belong? Nandrolone, testosterone, methyltestosterone, dihydrotestosterone and methandrostenolone.

29) Which is longer acting, GTN or ISMN?

30) Alendronate, risedronate and etidronate are members of which class of drugs?

31) What is miosis and what family of drugs is notorious for causing this?

32) Which enzymes are responsible for the initial breakdown of thyroid hormones?

33) For what is the drug simvastatin used?

34) What type of drug is actrapid?

35) What is the common name for fluoxetine?

36) What effect does amiodarone have on thyroid hormones?

37) Which of the following is not a medical emergency? Myocardial infarction, thyroid storm, malignant hypertension, hyperkalaemia and acute cranial arteritis.

38) Name three negative issues that may complicate the use of NSAIDS.

39) What is another name for *neuroleptics*?

40) A previously well 60 year old man develops a tremor, micrographia and clumsiness. Which drug would you test on the patient to confirm the suspected diagnosis?

# Session 4

# Multiple Choice Questions - Single Best Answers

91) Augmentin is

(A) Co-amoxiclav
(B) Cefuroxime
(C) Gentamicin
(D) Flucloxacillin and clavulanic acid
(E) Gentamicin and ampicillin

92) Atropine is an antagonist at

(A) $M_1$ receptors
(B) $M_2$ receptors
(C) $M_3$ receptors
(D) $M_1 - M_3$ receptors
(E) All muscarinic receptors

93) Aniline dye workers have an increased risk of

(A) Primary squamous cell carcinoma of the bladder.
(B) Primary transitional cell carcinoma of the bladder.
(C) Primary adenocarcinoma of the bladder.
(D) Small cell carcinoma of the bladder.
(E) Malignant melanoma of the bladder.

94) Which of the following statements is true of oxytocin?

(A) Oxytocin secretion is decreased during labour.
(B) In late pregnancy the uterus becomes insensitive to oxytocin.
(C) Oxytocin inhibits the milk ejection reflex.

(D) Oxytocin can be given to initiate labour.

(E) Oxytocin has an anti-ADH effect at the kidney.

95) A patient with symptomatic Wilson's disease and hepatic impairment is treated with penicillamine. The treatment is successful. Which of the following mechanisms accounts for the drug's success?

(A) Penicillamine acts as an antibiotic.

(B) Penicillamine binds to iron.

(C) Penicillamine binds to copper.

(D) Penicillamine binds to arsenic.

(E) Penicillamine is an antimetabolite.

96) A 33 year old man presents with malaise, generalized lymphadenopathy and tachypnoea.  He is pyrexial with a temperature of 38°C. His chest X-ray image shows bilateral mid-zone shadowing. Acutely, he is treated with co-trimoxazole, which resolves the respiratory signs and symptoms. What was the acute disorder with which the patient presented?

(A) Streptococcal pneumonia

(B) Staphylococcal lobar pneumonia

(C) Bronchiolitis

(D) Pneumocystis pneumonia

(E) Jirovecal pneumonia

97) Continuing the scenario above, the patient's underlying problems are managed by a cocktail of drugs including AZT. What is AZT?

(A) Adenozine triphosphate

(B) Zidovudine

(C) Azidothyroxine

(D) Zipovudine

(E) Adenine z-triphosphate

98) Which of the following is not true of AZT?

(A) It is an antimetabolite.

(B) It is a reverse transcriptase inhibitor.

(C) It is a nucleoside analogue.

(D) It inhibits viral RNA synthesis.

(E) It inhibits viral DNA synthesis.

99) A 45 year old motorcyclist is rushed into A and E after crashing into a car at 50 mph. Signs and symptoms of internal bleeding are noted. As part of supportive management gelofusine is given before surgery. What type of substance is gelofusine?

(A) A laxative

(B) An osmotic diuretic

(C) A colloid

(D) An antibiotic

(E) A sedative

100) Which of the following is not an ACE inhibitor?

(A) Verapamil

(B) Captopril

(C) Capoten

(D) Altace

(E) Vasotec

101) Which of the following is routinely used as a contrast agent in X-ray radiological investigations?

(A) Aluminium oxide
(B) Sodium nitrate
(C) Calcium sulphite
(D) Barium sulphate
(E) Magnesium sulphate

102) Chlorhexidine is a widely used drug in hospitals and in homes. Most doctors use it at least once a day. What class of drug is chlorhexidine?

(A) Diuretic
(B) Antiseptic
(C) Antispasmodic
(D) Antiarrhythmic
(E) Analgesic

103) A 60 year old woman suffers from chronic joint pain. The digits of her hand show ulnar deviation. She is a smoker and her cells exhibit HLA DR4. Her daughter has recently developed joint pain in her hands as well. Which of the following drugs would you use to manage the mother?

(A) Silver
(B) Oestrogen
(C) Sulphasalazine
(D) Interferon
(E) Paracetamol

104) At which receptor does actrapid exert its pharmacological effect?

(A) IGF-I receptor
(B) IGF-II receptor
(C) Insulin receptor
(D) EGF receptor
(E) PDGF receptor

105) Which of the following is true about the mechanism of action of GTN?

(A) It causes vasoconstriction.
(B) It is converted to the active form, nitrous oxide.
(C) It causes dilation of veins only.
(D) It leads to the stimulation of guanylate cyclase.
(E) It decreases intracellular cGMP levels in vascular smooth muscle.

106) Identify the established side-effect of ACE inhibitor use:

(A) Bradykinin depletion in the lungs.
(B) Hypokalaemia.
(C) Atrial fibrillation.
(D) Development of lymphoma.
(E) Precipitation or exacerbation of renal failure.

107) Which of the following drugs is an $\alpha$-adrenergic antagonist?

(A) Epinephrine
(B) Phenylephrine
(C) Prazosin
(D) Propranolol

(E) Atenolol

108) You find an individual collapsed and unconscious at the roadside. You arrive at the same time as the emergency medical personnel. The wife of the collapsed man tells you that he is prone to hypoglycaemia. You elect to manage the unconscious individual as if he were hypoglycaemic. Which of the following drug treatments would not be used?

(A) Oral glucose solution.
(B) Intravenous glucose solution.
(C) Intramuscular glucagon.
(D) Intravenous dextrose solution.
(E) Intravenous dextrose and saline.

109) A 65 year old male ex-bricklayer presents at Accident and Emergency complaining of new shortness of breath. He is a longstanding smoker. On examination the patient is tachypnoeic and febrile. Prior to this episode he had a productive cough continuously for five months. He has no history of symptomatic cardiac disease. He has no other personal medical history of note. You request a chest X-ray and send off blood cultures as well as taking blood for a full blood count, urea and electrolytes. At this point, what is the most likely working diagnosis?

(A) Pneumonia
(B) Chronic bronchitis
(C) Acute exacerbation of COPD
(D) COPD
(E) Acute exacerbation of asthma

110) Which of the following drugs would you not use to help the patient in the above acute scenario?

(A) Salmeterol
(B) Ipratropium bromide
(C) Oral prednisolone
(D) Amoxycillin
(E) Vancomycin

111) A 63 year old woman arrives at Accident and Emergency complaining of recent onset shortness of breath. There is no associated chest pain. She is an ex-smoker (20 pack years) with no other significant personal medical history. Her chest X-ray image shows a generalized fluffy infiltrate with prominent pulmonary vessels in the upper fields and peribronchial haze. Your impression is that this is pulmonary oedema. Which of the following is not a cause of pulmonary oedema?

(A) Vancomycin
(B) Levothyroxine
(C) Imiquimod
(D) Oestrogen
(E) Paracetamol

112) Considering the scenario above, how could frusemide (furosemide) help to manage the pulmonary oedema?

(A) Frusemide can reduce the fluid overload causing pulmonary oedema by increasing renal sodium and water excretion.
(B) It decreases the rate of coronary artery atherosclerosis.
(C) It acts on pulmonary arteries to cause vasoconstriction.

(D) It is a negative inotrope and chronotrope.

(E) Frusemide can reduce the fluid overload causing pulmonary oedema by increasing iron loss from the kidneys.

113) Assuming the pulmonary oedema originates in heart dysfunction, how might the β₁ agonist dobutamine, help in the management of pulmonary oedema?

(A) By stimulating dopamine receptors to increase noradrenaline release.

(B) By acting as a parasympathomimetic.

(C) By acting as a CNS neurotransmitter.

(D) By causing peripheral vasoconstriction to raise systemic blood pressure.

(E) By lessening heart failure through positive inotropic and chronotropic effects.

114) How does alendronate achieve its major therapeutic effect?

(A) It raises serum calcium ion concentrations.

(B) It stimulates vitamin D action.

(C) It increases calcium absorption from the GI tract.

(D) It inhibits osteoclast mediated bone resorption.

(E) It stimulates osteoblast action and bone deposition.

115) A previously well 25 year old medical student went on holiday to Ghana in West Africa. He was careful to take all of the appropriate vaccinations before departing to Africa. The student also took oral chloroquine commencing one week before departure and ceasing four weeks after he returned. Six weeks after returning from his holiday he attended his GP surgery complaining of a bad flu. On questioning by

his GP, the student admitted that he had an intermittent fever, which seemed to be worse every third day. On examination, hepatomegaly and splenomegaly were noted. No lymphadenopathy was found. The General Practitioner took blood samples that showed anaemia and hyperbilirubinaemia. Why did the medical student become unwell?

(A) Because of protozoal drug resistance.
(B) Because he failed to obtain the appropriate vaccination.
(C) Because he wore long-sleeved clothes.
(D) Because he had unsafe sex with a local prostitute.
(E) Because he drank dirty water.

116) As a junior doctor working in Neurology you admit a patient with suspected herpes simplex encephalitis. Which of the following drugs would you give the patient in the acute scenario?

(A) HAART
(B) Gentamicin
(C) Ampicillin
(D) Prednisolone
(E) Acyclovir

117) Which of the following analgesic regimes is most likely to be used in the management of mild pain?

(A) 10mg morphine prn
(B) 1g paracetamol qds po
(C) 30mg codeine qds po
(D) 1g paracetamol qds po and 75mg diclofenac bd po
(E) Co-codamol and diclofenac

118) A 37 year old man presents at his General Practitioner's surgery complaining of a widespread red rash that is itchy, swollen and crusting. There is no obvious pattern to the distribution of the rash. There is no pyrexia or systemic symptomatology. He has no other significant past medical history. Despite an extensive history taking, no precipitating factor is identified. His mother has asthma and his father suffers from hayfever. You make a working diagnosis. Which of the following would not be part of the management?

(A) Discussion, education and reassurance.
(B) Topical hydrocortisone.
(C) Emollients (moisturizing cream).
(D) Calamine lotion.
(E) Ultraviolet light.

119) Which of the following drugs can be used to most successfully treat an MRSA infection? (MRSA = Methicillin Resistant Staphylococcus Aureus)

A) Benzylpenicillin
B) Cephalexin
C) Amoxicillin
D) Vancomycin
E) Lancomycin

120) Which of the following substances is not used to manage pre-eclampsia?

(A) Magnesium sulphate
(B) Nifedipine
(C) Labetalol

(D) Nicardipine

(E) Calcium sulphate


# Short Answer Questions:

## Question 7

A 70 year old male attends his GP's surgery. This patient is usually fit and well. He has mild osteoarthritis of his right knee, otherwise he has no medical history of note. Currently he complains of a recent onset severe headache and temporal pain, which is exacerbated when he combs his hair. He thinks that his vision has recently deteriorated on the same side that he experiences the pain and headaches.

A) What is the most likely diagnosis? (1 mark)



B) What blood test results should you see before initiating treatment? (1 mark)



C) What is the standard treatment for your diagnosis? (2 marks)

D) Name four side-effects of long term use of this class of drug. (4 marks)

E) Name three hyperglycaemic hormones. (3 marks)

F) This unfortunate patient develops type 2 diabetes mellitus. He requires insulin for the management of his diabetes mellitus. Name one short acting and one long acting insulin or insulin analogue. (2 marks)

G) This patient continues to be unfortunate and develops an allergy to one of his insulin preparations. He suffers anaphylaxis in front of you. What drug would you use in his immediate management? What concentration/dilution would you use and how would you administer it? (3 marks)

H) After taking a more careful history from this patient you note that he is atopic. He suffers from hayfever in the summer time and takes nasal Beconase and Benadryl. What are these drugs and what are their mechanisms of action? (4 marks)

## Viva Voce Questions:

41) Aspirin, diclofenac and ibuprofen belong to which group of drugs?

42) For what is carbimazole used?

43) Which radioactive drug can be used to treat hyperthyroidism? What notorious risk is associated with this drug?

44) Which receptors does labetalol block?

45) Which has greater bioavailability, a sublingual drug or an intravenous drug?

46) Is warfarin normally predominantly protein bound or non-protein bound in plasma?

47) What class of antibiotics are vancomycin and teicoplanin?

48) Name three statins.

49) What type of drug is propranolol and what is its modern equivalent?

50) In liver failure does the free circulating level of a patient's phenytoin rise or fall?

51) Do children metabolize hepatically degraded drugs faster or more slowly than adults? Why?

52) Phenytoin, carbamazepine and sodium valproate are anticonvulsants. What common mechanism of action do they share?

53) How does imipramine work?

54) Name three classes of drugs that you can use to treat depression.

# Answers

# Session 1

# Multiple Choice Questions
## Answers

| | | |
|---|---|---|
| 1) B | 11) E | 21) E |
| 2) D | 12) C | 22) B |
| 3) E | 13) D | 23) A |
| 4) A | 14) D | 24) A |
| 5) E | 15) B | 25) B |
| 6) C | 16) D | 26) E |
| 7) B | 17) A | 27) C |
| 8) E | 18) B | 28) C |
| 9) C | 19) C | 29) B |
| 10) D | 20) B | 30) B |

1) A normally fit and well 75 year old woman presents at A and E "off legs." She is confused and has a temperature of 37.9°C. She has no signs of trauma, no neurological signs, her blood glucose is within normal limits and so are her thyroid function tests. Which of the following are you most likely to prescribe?

A) A cephalosporin
B) 200mg trimethoprim bd po
C) 50mg atenolol od po
D) 10 units of insulin iv
E) 1g paracetamol qds po

1) **B**

**200mg trimethoprim bd po**. This is a standard treatment regime for a urinary tract infection. "Off legs" means that the patient is unable to cope alone, although she has previously been managing well. The differential diagnosis for such a situation includes dehydration or hypovolemia, hypothyroidism, UTI and stroke/TIA. These diagnoses can usually be made and managed initially by the junior doctor.

Cephalosporins are often used for sepsis, they are not specific for the Gram negative bacteria that usually cause UTI. Atenolol is a beta blocker usually used for hypertension and 10 units of intravenous insulin is the beginning of the management for DKA (or hyperkalaemia). Paracetamol is usually prescribed for pyrexia or pain.

2) Which of the following analgesics does not contain codeine?
(A) Co-codamol
(B) Tylenol
(C) Nurofen plus
(D) Anadin extra
(E) Solpadeine max

2) **D**

All of the drugs listed contain codeine except **anadin extra**; the latter contains paracetamol, aspirin and caffeine:

Co-codamol  (paracetamol plus codeine)
Tylenol (acetaminophen plus codeine)
Nurofen plus (ibuprofen plus codeine)
Solpadeine max (paracetamol plus codeine)

Note that in the USA *acetaminophen* is the term used instead of *paracetamol*.

3) Which of the following is not a side-effect of opiate use?
(A) Constipation
(B) Respiratory depression
(C) Addiction
(D) Mood elevation
(E) Diplopia

3) **E**

Opiates are notorious for causing constipation, respiratory depression and mood elevation.  Susceptible individuals have an increased risk of addiction to opioids. Opiates do not usually cause **diplopia**.

4) Which of the following antibiotics does not achieve its therapeutic

effect through the inhibition of cell wall synthesis?

(A) Quinolones

(B) Carbapenems

(C) Vancomycin

(D) Cephalosporins

(E) Penicillins

4) **A**

**Quinolones** inhibit nucleic acid synthesis in the affected microorganisms. The other antibiotics all inhibit microbial cell wall synthesis. Pharmacologically, it is useful for the physician to know when they are switching classes of antibiotics, so that effective combinations can be used to treat the infection.

5) You are called to a ward where an agitated elderly man is physically threatening patients and nurses. The security staff have already been called. The lead nurse asks for help in sedating the patient. Which of the following prescriptions is most helpful in this situation?

A) 5mg haloperidol po

B) 40mg diazepam po

C) 3mg atropine iv

D) 1:1000 adrenaline

E) 2mg haloperidol im

5) **E**

This agitated patient needs to be sedated by a drug – classically antipsychotics (such as haloperidol or droperidol) or benzodiazepines (such as midazolam) are used. The patient is unlikely to cooperate with an oral dose and in any case **intramuscular** administration of haloperidol is the preferred route for *rapid action*. Initially, doses under 10mg of haloperidol are used, with the amount being increased with each injection until effective. An initial dose intramuscular dose of **2mg** in an elderly man would not be unusual. 40mg of diazepam is too high a dose of this benzodiazepine. Atropine is not a sedative and the dose suggested is the amount used in cardiac arrests. The adrenaline dose is the amount used in anaphylaxis. Adrenaline is not a sedative.

6) A 55 year old woman with a long history of intractable depression is prescribed selegiline to which she responds well. She has no other significant comorbidity or family history. She attends her daughter's wedding, which she plays a major role in organizing. During the wedding she drinks and eats heartily. In the middle of the wedding she complains of chest pain, palpitations, dizziness and headache. Her nose starts to bleed. She becomes drowsy and confused. What is the most likely diagnosis?

A) DVT and PE

B) MI

C) MAOI reaction

D) Malignant hyperthermia

E) Alcoholism

6) **C**

The patient is taking selegiline which is a **m**ono **a**mine **o**xidase **i**nhibitor used in the treatment of depression. She has been drinking and is at a wedding where she may have inadvertently ingested a tyramine containing food that reacts with the MAOI to cause a hypertensive crisis ("cheese effect"). MAOIs inhibit tyramine breakdown leading to tyramine accumulation which is believed to cause noradrenaline release and thus the hypertensive crisis. The patient was most likely to be experiencing an **MAOI reaction**. The nose bleeding and headache, together with her medication, should make the reader suspicious of an acute hypertensive episode. Considering her age, a myocardial infarction should be on the differential diagnosis, but she has no personal or family history to suggest an MI and the pattern of signs and symptoms fit better with an MAOI effect. There is no evidence of raised body temperature. There are no signs or symptoms specific for a DVT. Alcoholism is not part of this patient's history, nor does it account for her signs and symptoms.

7) In the above scenario the doctor arriving on the scene makes the correct diagnosis and in the acute setting would initially prescribe which of the following?

A) 300mg aspirin po
B) 10mg labetalol iv
C) 40mg nifedipine po
D) Oxygen
E) Sublingual GTN

7) **B**

Any of the following drugs can be used in the treatment of hypertensive crises:

Sodium nitroprusside

Labetalol

Calcium channel antagonists (e.g. nifedipine)

These drugs are given intravenously for rapid effect.

In this example, **10mg labetalol iv** would be a good initial treatment. The suggested initial nifedipine dose (C) is too high, has been prescribed for oral administration (which may require more patient cooperation than is available) and is not the most rapid means of administration. The GTN, aspirin and oxygen relate to initial treatment of *cardiac* chest pain.

8) In clinical pharmacological terms, which of the following is the odd one out?

A) Paricalcitol
B) Calcitriol
C) Alfacalcidol
D) Cholecalciferol
E) Cholesterol

8) **E**

**Cholesterol** is the odd one out because options A-D are all pharmacological variants of vitamin D.

9) You have a patient that is apparently thyrotoxic. On examination a goitre is obvious. You submit venous blood to test for T4 and TSH levels. The blood tests show significantly elevated T4 and significantly diminished circulating TSH. What is the most likely diagnosis?

A) Primary hypothyroidism
B) Secondary hypothyroidism
C) Primary hyperthyroidism
D) Secondary hyperthyroidism
E) Secondary hyperparathyroidism

9) C

The diagnosis is **primary hyperthyroidism**. A *thyrotoxic* patient demonstrates signs of supraphysiological thyroid hormone levels. Usually the underlying cause is hyperthyroidism, however an over-administration of thyroid hormones is also a possibility. In this case the patient has a large thyroid gland together with the thyrotoxicosis – suggesting a primary lesion in the thyroid that is secreting thyroid hormones and causing subsequent negative feedback to decrease TSH levels. (The patient is most likely to be suffering from Graves disease, a toxic multinodular goitre or a toxic thyroid adenoma).

10) The patient described in the previous scenario declines surgery and so a drug treatment is started. Which of the following is most likely to be successful?

A) 0.9mg TSH bd iv
B) 100 intl units calcitonin od sc
C) 10mg salbutamol tds
D) 10mg carbimazole od po
E) 25 micrograms T4 po

10) D

**Carbimazole** inhibits thyroperoxidase, an enzyme responsible for the biosynthesis of thyroid hormones. The maintenance dose of carbimazole is approximately 5-15mg per day. β adrenergic *antagonists* can be used in the treatment of hyperthyroidism (*not agonists* such as salbutamol). T4 supplementation would exacerbate hyperthyroidism, as would TSH supplementation. Calcitonin acts to lower the plasma calcium ion concentration.

11) You are a GP managing a 48 year old diabetic patient. The patient has a total blood cholesterol of 7 mmol/l, a resting blood pressure of 150/95 and an HbA1c > 9%. You elect to start the patient on a statin. Which of the prescriptions below would be most helpful clinically?
A) 40mg pravastin od iv
B) 40mg pravastatin qds po
C) 2.5mg pravastatin od po
D) 20mg simvastatin qds po
E) 40mg simvastatin od po

11) **E**

**40mg simvastatin od po** is a standard therapeutic dose. The normal dose range of simvastatin is 5 – 80mg daily as an oral drug. The patient is more likely to comply with the simpler dosing regime of one oral tablet daily. The normal dose of pravastatin is approximately 10 – 80mg daily. The drug is usually given orally - intravenous administration would be difficult to manage in the community.

*JBS 2 Study: indications for statin therapy in type 1 or type 2 diabetes:*
- Age > 40 years.
- Retinopathy of greater than background severity.
- Nephropathy, including microalbuminuria.
- Poor glycaemic control (HbA1c > 9%).
- Hypertension requiring treatment.
- Elevated total cholesterol (> 6.0 mmol/l).
- Metabolic syndrome.
- Family history of CHD in a first-degree relative.

12) Tamoxifen is

A) An oestrogen receptor agonist.

B) An oestrogen receptor antagonist.

C) A mixed oestrogen receptor agonist and antagonist.

D) A progesterone receptor agonist.

E) A progesterone receptor antagonist.

12) **C**

Tamoxifen is a **mixed oestrogen receptor agonist and antagonist**; the agonist activity accounts for the increased risk of oestrogenically driven endometrial carcinoma.

13) You are a junior doctor called to a medical ward where a 40 year old woman has been suffering from hiccups for the last two days. It is night time and the attending nurse would like you to prescribe something to control the hiccups so that the patient can sleep. Which of the following is most likely to be an effective prescription?

A) 1 litre of normal saline iv

B) 1g paracetamol qds po

C) 20mg buscopan (hyoscine butylbromide) qds po

D) 50mg chlorpromazine tds po

E) 50mg voltarol (diclofenac) tds po

13) **D**

**Chlorpromazine** is used to treat intractable hiccups (that have persisted for at least 48 hours). **50mg tds po** is a therapeutic dose. Voltarol (diclofenac) is an NSAID analgesic, paracetamol is an analgesic and antipyrexial and buscopan is an antispasmodic. Normal saline is not a treatment for hiccups. It is not clear how chlorpromazine achieves its effect.

14) Which of the following substances can be used to manage constipation?

A) Lactose

B) Glucose

C) Galactose

D) Lactulose

E) Sucrose

14) **D**

**Lactulose** is an osmotic laxative used in the management of constipation. It is a disaccharide of fructose and galactose. In the GI tract lactulose ferments into lactic acid and acetic acid. The latter are the osmotically active substances.

15) A patient with chronic mental health problems has been taking medication for a decade. He has recently developed tardive dyskinesia and oculogyric crises. What class of drug is this patient likely to have been taking?

A) Monoamine oxidase inhibitor

B) Antipsychotic

C) Serotonin specific re-uptake inhibitor

D) Benzodiazepine

E) Opiate

15) **B**

Tardive dyskinesia and oculogyric crises are classical complications of long term **antipsychotic** (neuroleptic) use. The other classes of drugs are not known for these complications.

16) Which of the following drugs is correctly matched with its mechanism of action?

A) Atropine; antagonist at nicotinic receptors.

B) Valium; antagonist at GABA$_A$ receptors.

C) Frusemide; osmotic diuretic.

D) Neostigmine; reversible acetylcholinesterase inhibitor.

E) Nifedipine; sodium channel antagonist.

16) **D**

Neostigmine is used in the treatment of myasthenia gravis because it **reversibly inhibits acetylcholinesterase** to raise acetylcholine levels at the neuromuscular junction. Atropine is an antagonist at muscarinic receptors. Valium (diazepam) is an agonist at GABA$_A$ receptors that facilitates GABA binding. Frusemide is a loop diuretic. Nifedipine is a calcium channel antagonist.

17) Which one of the following drugs has not been used to manage hypertension?

A) Aspirin

B) Frusemide

C) Captopril

D) Atenolol

E) Verapamil

17) **A**

**Aspirin** is used as an analgesic and to reduce the risk of blood clots but it has no direct role in the management of hypertension. The classes of drugs that have had a role in the management of hypertension include:

Diuretics e.g. frusemide

Beta adrenergic receptor antagonists e.g. atenolol

Alpha adrenergic receptor antagonists

Calcium channel blockers e.g. verapamil

ACE inhibitors e.g. captopril

Angiotensin II receptor blockers

Direct renin inhibitors

18) Which of the following is a contraindication for the use of streptokinase?

A) Ischaemic stroke within the previous 18 months.

B) Pregnancy

C) Caput medusae

D) Gastritis

E) Renal failure

18) **B**

**Pregnancy** is a contraindication for the use of streptokinase – the risk of a haemorrhagic event because of the vascularity of the placenta and fetal blood flow is considered too high. An ischaemic stroke 18 months previously is not a contraindication as this is not evidence of a predisposition to bleed. Caput medusae may indicate hepatic disease but it is the presence of varices, which may rupture and bleed catastrophically, that is the more significant risk. Gastritis is very common and has little to do with a bleeding tendency. However a history of peptic ulceration would be worrying. Renal failure does not predispose to significant bleeding.

**The following are contraindications because of the risk of serious bleeding with streptokinase:**
• previous haemorrhagic stroke
• any stroke within 6 months
• recent arterial/other major surgery
• peptic ulceration/internal bleeding
• oesophageal varices
• pregnancy
• severe proliferative retinopathy

19) How do thiazides work? By inhibiting

A) The $Na^+2Cl^-K^+$ transporter on the luminal side of the ascending loop of Henle.

B) The $Na^+/K^+$ pump on the basolateral side of the ascending loop of Henle.

C) The $Na^+Cl^-$ contransporter on the luminal side of the distal tubule.

D) Carbonic anhydrase.

E) Creatine kinase.

19) **C**

Thiazides are distal tubule diuretics that bind to the chloride site on the **sodium chloride cotransporter** and **inhibit** it. Sodium is not reabsorbed so a natriuresis occurs. Water then follows the sodium. Option (A) is the mechanism of action of loop diuretics.

20) It is 9pm and you are a junior doctor called to see a 35 year old male inpatient who is claiming to have a severe headache with associated nausea, vomiting and photophobia. He has been an inpatient for three weeks for the management of inflammatory bowel disease. He had a lunch that included oranges and chocolate. The headache has persisted for 7 hours despite his nurse giving him paracetamol. His blood pressure and pulse are normal. On examination no signs of meningism are elicited and no fever is noted. What is the most likely diagnosis?

A) Inflammatory bowel disease

B) Migraine

C) Tension headache

D) Meningitis

E) Encephalitis

20) **B**

The most likely diagnosis is a migraine; this can be triggered by factors including citrus fruit, caffeine, chocolate, cheese, the contraceptive pill and alcohol. The diagnosis is consistent with the severe headache, nausea, vomiting and photophobia. The lack of signs of meningism and the fact that the patient is a longstanding inpatient makes meningitis/encephalitis less likely. (A complete history would have looked for evidence of a prodrome with visual disturbances or an "aura"). A tension headache should be on the differential diagnosis but is not usually associated with photophobia.

21) What drug could you use to treat the patient in the previous question?

A) Paracetamol

B) Labetalol

C) Oestrogen

D) Clozapine

E) Sumatriptan

21) **E**

Sumatriptan is an effective drug in the treatment of migraine. A typical dose is 50mg po od. It is known to work by antagonizing 5-hydroxytryptamine function. Excess CNS 5-hydroxytramine is believed to cause migraines. Ergotamine, beta blockers and methylsergide have also been used to treat migraines. Labetalol is a mixed alpha and beta blocker that is inappropriate for the treatment of migraines.

22) At cardiac arrest situations what dose of atropine is recommended?

A) 1mg

B) 3mg

C) 5mg

D) 10mg

E) 20mg

22) **B**

3mg of atropine is recommended; this muscarinic receptor antagonist blocks the vagal inhibition of the heart rate and hence can reverse bradycardias.

23) You are called to a ward at 1 am because a patient is having difficulty sleeping. The patient has a history of insomnia with no other relevant comorbidity. What could you prescribe to help the patient sleep?

A) 10mg temazepam od po (nocte)

B) 400mg clozapine od po (nocte)

C) 5mg haloperidol od im (nocte)

D) 1mg risperidone bd po (nocte)

E) 10mg diazepam po od (nocte)

23) **A**

Short to medium term acting benzodiazepines are used as sedatives to help with sleeplessness. **Temazepam** is such a benzodiazepam. Diazepam is inappropriate because it acts over a longer time and is more likely to cause daytime drowsiness.

Clozapine, risperidone and haloperidol are antipsychotic (neuroleptic) medications that may be used for emergency sedation but not insomnia.

24) As a junior doctor you are called to see a patient who has developed new musculoskeletal back pain, partly because of the extensive period of bed rest. What prescription would you try initially?

A) 400mg ibuprofen tds po

B) 10mg morphine od po

C) 50mg diclofenac od po

D) 1 tablet cocodamol qds po

E) 20mg morphine od po

24) **A**

The maximum recommended therapeutic dose of **ibuprofen** in a 24 hour period is approximately 3200mg, usually given in 4 – 6 hour dosing regimes. Hence **400mg** three times a day is an acceptable therapeutic regime. Ibuprofen is a first line treatment for musculoskeletal pain. Diclofenac is too powerful an NSAID for first line treatment. Opiate containing analgesics are also inappropriate in the initial management of musculoskeletal pain.

25) Which of the following is used as a general anaesthetic?

A) Diazepam

B) Halothane

C) Lignocaine

D) Procaine

E) L-dopa

25) **B**

**Halothane** is used as a general anaesthetic. Diazepam is a benzodiazepine sedative. Lignocaine and procaine are local anaesthetics. L-dopa is used to treat Parkinson's disease.

26) Which of the following is not a potential side-effect of (or adverse reaction to) general anaesthetic use?

A) Malignant hyperthermia

B) Myocardial infarction

C) Aspiration
D) Hepatitis
E) Pancreatitis
26) **E**
**Pancreatitis** is not a recognized complication of general anaesthetic use. By lowering systemic blood pressure general anaesthesia can cause myocardial infarction. This is the leading cause of death under general anaesthesia. Because reflexes are suppressed by general anaesthesia, aspiration and aspiration pneumonia are not uncommon complications. Hepatitis is a rare complication of halothane use (the hepatitis believed to be either because of toxic metabolites or because of an immunological reaction). Malignant hyperthermia is a rare event in which a rapid rise in body temperature occurs in reaction to a general anaesthetic – it is a medical emergency.

27) A patient has the following plasma concentrations; $[Na^+]$ = 120mM, $[K^+]$ = 4.7mM, $[Ca^+]$ = 2.5mM and pH = 7.3. The patient is
A) Hyperkalaemic and acidotic
B) Normokalaemic and alkalotic
C) Hyponatraemic and acidotic
D) Hypercalcaemic and hypokalaemic
E) Hyperkalaemic and alkalotic
27) **C**
The patient is **hyponatraemic and acidotic**.
Normal ranges:
$[Ca^+]$ = 2.25 – 2.75mM
$[K^+]$ = 3.5 – 5mM
$[Na^+]$ = 135 – 145mM
pH = 7.35 – 7.45

28) A patient with severe aortic stenosis is given a mechanical heart valve replacement. Which lifelong drug will the patient require?
A) Prednisolone

B) Benzylpenicillin

C) Warfarin

D) Methotrexate

E) Aspirin

28) **C**

**Warfarin** is used to anticoagulate the blood of a patient with a mechanical heart valve to achieve an appropriate range of INR (International Normalized Ratio) by titrating the drug dose.

29) After a myocardial infarction a patient is put on regular aspirin doses. What is the prescription most likely to be?

A) 35mg od po

B) 75mg od po

C) 150mg od po

D) 225mg od po

E) 400mg od po

29) **B**

The standard post-MI maintenance dose of aspirin is **75mg od po**.

| Post-MI Management: |
| --- |
| Reassurance |
| Oxygen |
| Morphine |
| Aspirin |
| Nitrates |
| Clopidogrel |
| Enoxaparin |

30) A 45 year old patient with HIV complains about a burning sensation in his mouth. On examination raised white plaques are evident. What drug would you use to manage the condition?

A) Cefuroxime

B) Nystatin

C) Penicillin

D) Doxycycline

E) Chloramphenicol

30) **B**

This individual with HIV is likely to be relatively immunosuppressed and prone to oral fungal (candidal) infection. The standard treatment for this is oral **nystatin**. None of the other drugs are anti-fungals. Nystatin is a polyene antifungal that binds to ergosterol in the cell membrane. Toxicity is caused by an alteration in the cell wall permeability resulting in leakage of cell contents.

# Short Answer Questions:
# Answers

## Question 1

A 25 year old woman presents at her GP surgery with a 24 hour history of a very painful left knee. She gives no history of trauma. She has no family history of rheumatoid arthritis. 48 hours ago she was at a wedding in Edinburgh. She has no other medical history or associated symptoms of note.

A) What are your three potential diagnoses (differential diagnosis) for acute severe joint pain in such a person? (3 marks)

**Septic arthritis** occurs most commonly in children/the young. (1 mark)

**Gout** has a dietary trigger. (1 mark)

**Rheumatoid arthritis** is more common in women and is autoimmune.

(1 mark)

(0.5 mark for **Pseudogout**)

B) A fine needle aspirate of the joint is taken and sent to microbiology. As a precaution antibiotic cover is started. If the most likely infective agent is a skin staphylococcus, which family of antibiotics would you prescribe first? (1 mark)

**Penicillin** is a readily available antibiotic that is effective against Gram positive bacteria such as staphylococcus aureus.

C) Eventually the results come back and indicate that the aspirate was sterile and without neutrophils – there is no evidence of septic arthritis. You consider gout as a cause of the young woman's symptoms. What substance deposited in the joint is directly responsible for causing the joint pain? (1 mark)

**Uric acid** (as crystals) (1 mark)

D) What would you prescribe as an analgesic initially? What would you prescribe to decrease the chances of future acute episodes? (2 marks)

**Indomethacin** or **ibuprofen.** (1 mark) An NSAID can be used as an acute analgesic.
**Allopurinol** to prevent future episodes. (1 mark) It inhibits xanthine oxidase, responsible for synthesizing uric acid. Decreasing the synthesis of uric acid decreases the risk of further attacks.

E) Before prescribing NSAIDS to a patient, what three questions should the patient be asked? (3 marks)

i) **Does the patient have a history of peptic ulceration?** NSAIDS can cause inhibition of cyclo-oxygenase in the stomach and impair healing, so increasing the chances of ulceration and a catastrophic haemorrhage.

ii) **Does the patient have asthma?** NSAIDS can exacerbate asthma.

iii) **Does the patient have kidney disease?** NSAIDS can cause renal damage (e.g. interstitial nephritis) and subsequent impairment.

F) After being prescribed an NSAID the patient experienced oliguria and subsequent blood tests indicated renal impairment. The potassium ion concentration rose to 7.5mM. What is the normal range of serum potassium concentrations and what is the clinical significance of 7.5mM? (2 marks)

**The normal range is approximately 3.5 – 5 mM.**
7.5mM is high enough to be a **medical emergency.**

G) What is the immediate drug management of hyperkalaemia and what is the function of each drug? (8 marks)

1) 10ml 10% **calcium gluconate** iv – protects **cardiac cell membranes.**
2) 50ml 50% **glucose** iv plus 15 units actrapid insulin – shifts **potassium** into cells.
3) 30g **calcium resonium** per rectum – **removes potassium** from body.
4) 10mg nebulised **salbutamol** – shifts **potassium into cells.**


## Question 2

A previously well 37 year old motorcyclist is involved in a serious road traffic accident. On examination it is noted that his abdomen is enlarged and shows generalized tenderness. His pulse is 120 bpm and regular - his blood pressure is 80/40mmHg. Abdominal and pelvic X-ray images indicate that he has suffered three fractured ribs and a fractured pelvis.

A) What is the term for his current state and what is the most likely underlying cause? (2 marks)

Cardiovascular **shock** (1 mark)
**Intra-abdominal haemorrhage** (1 mark)

B) Which two classes of fluid can be used to replace his cardiovascular losses? (2 marks)

**Colloid** (1 mark)
**Crystalloid** (1 mark)

C) Give an example of each class. (2 marks)

Colloid – **Blood**, Haemaccel, Gelofusine, Hydroxyethyl starch (any for 1 mark)
Crystalloid – **Saline**, Ringer's lactate, 5% dextrose (D5W) (any for 1 mark)

D) Which class of fluid provides the better prognosis in this and similar situations? (1 mark)

**Neither is more effective at improving prognosis**. A larger volume of crystalloid can have the same effect as a smaller volume of colloid in maintaining circulatory volume. In practice crystalloids are more commonly used because of convenience and cost. (1 mark)

E) As a result of this trauma the patient requires major surgery. Which covering antibiotics would you favour for such surgery? (2 marks)

Major abdominal surgery often uses the combination of **cefuroxime** and **metronidazole**. This provides effective cover against obligate anaerobes (metronidazole) and Gram negative bacteria (cefuroxime). Ceftriazone/Penicillin is also often used.

F) As part of the surgery it is necessary to remove the patient's spleen. Which drug or class of drug might the patient have to take for the rest of his life? (1 mark)

**Penicillin is commonly prescribed.**

G) Three days after the surgery the patient complains of left knee pain that is resistant to the post-operative analgesics that he is taking. On examination the left knee is extremely tender, warm, swollen and red. On admission the left knee appeared undamaged by the accident and the patient did not complain of pain. The patient has no history of rheumatoid arthritis and no previous trauma to the knee.
A fine needle aspirate is taken and the fluid sent to microbiology, biochemistry and cytopathology. The presence of uric acid crystals is confirmed.

What is the joint disease that this unfortunate patient is experiencing and what is the likely trigger? (2 marks)

**Gout; the trauma associated with major surgery is the likely trigger.** (Aspirin may trigger gout but it is unlikely that this analgesic would have been used after his surgery).

H) Name two drugs that can be used in the acute management of gout and briefly describe their mechanisms of action. (4 marks)

**Indomethacin or ibuprofen**. Non-selective cyclo-oxygenase inhibition preventing synthesis of inflammatory mediators by inhibiting the conversion of arachidonic acid to prostaglandins and thromboxane.

**Colchicine**. Inhibits microtubule formation and so it inhibits neutrophil motility and activity to produce an anti-inflammatory effect.

**Glucocorticoids**. Anti-inflammatory action by diffusing through the cell membrane and binding to steroid receptors. The steroid hormone receptor complex diffuses to the nucleus to alter transcription of key proteins important in inflammation.

I) Name two classes of drugs that can be used to prevent acute attacks of gout and give an example of each. (4 marks)

**Xanthine oxidase inhibitor** (inhibiting biosynthesis of uric acid) e.g. allopurinol or febuxostat.
**Uricosuric** (increasing excretion of uric acid) e.g. probenecid or sulfinpyrazone.

# Viva Voce Questions
# Answers

1) A patient has had a recent clinically confirmed myocardial infarction. The patient is not suitable for angioplasty – what drug treatments are possible?

Commonly streptokinase or tissue plasminogen activator (tPA) are used.

2) ACE inhibitors are now commonly prescribed medicines. What should an alert doctor be worried about when using these drugs?

a) Hyperkalaemia  b) Cough/angioedema  c) Presence of renal artery stenosis.

3) Name a potassium sparing diuretic.

Spironolactone, amiloride and triamterene are common potassium sparing diuretics.

4) Name two organic nitrates used in the management of cardiac disease.

Isosorbide mononitrate, Glyceryl trinitrate, Isosorbide dinitrate.

5) What drug would you use to treat *status epilepticus* acutely?

Lorazepam or diazepam are commonly used.

6) What abbreviation do you use if you want a drug to be given at once?

STAT (from the latin *statim* meaning **immediately**).

7) What does the term tocolytic mean?

It is a drug used to suppress a (premature) labour. The $\beta_2$ agonist ritodrine is an example of such a drug.

8) What class of receptor does oxytocin bind to and stimulate physiologically?

The oxytocin receptor is one of the **G-protein coupled receptor** family.

9) Which of the following drugs is most likely to help you pass a pharmacology exam; LSD, Ecstasy, Amphetamine, Heroin or Cocaine?

Amphetamine is commonly used by students because of its stimulant action to cause greater alertness; it is often used for cramming sessions.

10) Name 5 classes of diuretics.

Any five of:

Loop diuretics, thiazides, aldosterone antagonists, epithelial sodium channel blockers, carbonic anhydrase inhibitors, osmotic diuretics and ADH antagonists.

11) Name a first line drug used in the treatment of cystitis.

Trimethoprim or nitrofurantoin.

12) You are on an ophthalmology rotation and you want to dilate a patient's pupil to examine the retina fully. Classically which drug is generally used?

Atropine.

13) Do glucorticoids preferentially decrease the frequency or intensity of asthma attacks?

A glucocorticoid's greatest effect is in decreasing the *frequency* of asthma attacks.

# Session 2

# Multiple Choice Questions
## Answers

| | | |
|---|---|---|
| 31) E | 41) C | 51) E |
| 32) D | 42) A | 52) B |
| 33) A | 43) D | 53) A |
| 34) E | 44) C | 54) A |
| 35) D | 45) D | 55) E |
| 36) B | 46) D | 56) A |
| 37) A | 47) B | 57) D |
| 38) B | 48) A | 58) C |
| 39) D | 49) D | 59) A |
| 40) D | 50) C | 60) A |

31) A 20 year old female patient is an emergency admission to hospital after having taken 2 bottles of paracetamol and written a suicide note; the events happening within the last 6 hours. Which organ is most susceptible to acute toxicity?
A) Kidney
B) Heart
C) Pancreas
D) Spleen
E) Liver
**31) E**
Hepatic necrosis is a major adverse effect of a paracetamol overdose. Drugs such as N-acetylcysteine are given in an attempt to protect the **liver**.
32) Considering the scenario in the previous question, what drug would you wish to prescribe to counteract the paracetamol overdose?
A) Omeprazole
B) Cimetidine

C) Naloxone

D) N-acetylcysteine

E) Lansoprazole

32) **D**

**N-acetylcysteine** can be given initially as 150mg/kg in 200ml 5% glucose over 15 minutes. A metabolite of paracetamol is responsible for the hepatotoxicity by depleting the stores of the antioxidant glutathione. Naloxone is used to treat opiate poisoning.

33) An emaciated 50 year old man is found collapsed in his flat on a Saturday evening. No one had seen him since the previous Friday. He has a social history that includes unprotected sex with multiple partners and recreational drug use. On examination of the patient you notice needle marks on his forearms and feet. The patient's breathing is slow and shallow and his pulse is 50 bpm. His pupils have contracted to pin points bilaterally (miosis). Which of the drugs below would be most likely to result in a positive clinical outcome?

A) Naloxone

B) Codeine

C) Tramadol

D) Fentanyl

E) N-acetylcysteine

33) **A**

The pin point pupils (miosis), decreased level of consciousness and slow and shallow breathing are suggestive of an opiate overdose. Opiates are notorious respiratory depressants that, if left unopposed, can cause death. The classic treatment is naloxone administration; it is injected intravenously for most rapid effect. However it can also be administered intramuscularly or subcutaneously. **Naloxone** acts in the central nervous system as a high affinity competitive antagonist of μ-opioid receptors. Hence it directly counteracts the opiate overdose to cause a rapid withdrawal. The μ-opioid receptors mediate analgesia, euphoria, miosis and physical dependence.

Codeine, tramadol and fentanyl are opiates that would exacerbate the patient's condition. N-acetylcysteine is used for overdoses of paracetamol.

34) Which of the following is not a compound that is used to treat iron deficiency anaemia?

A) Iron fumarate

B) Iron citrate

C) Iron sulphate

D) Iron gluconate

E) Iron nitrate

34) **E**

**Iron nitrate** is not used to treat iron deficiency anaemia. A commonly used medication is iron sulphate. However, this can cause:

- nausea
- abdominal pain
- heartburn
- constipation
- diarrhoea
- black stools

Iron fumarate, iron citrate and iron gluconate are alternative pharmacological sources of iron. These are better tolerated by patients who believe that they are "allergic" to iron supplements because of the gastrointestinal irritation caused by iron sulphate.

35) A patient with epigastric pain is investigated by endoscopy and biopsies are taken. On microscopic examination the histological features of a Helicobacter Pylori infection are noted. You elect to treat the patient with two different antibiotics and a proton pump inhibitor, omeprazole. What dose of omeprazole would you prescribe?

A) 2.5mg od po

B) 5mg od po

C) 10mg od po

D) 20mg od po

E) 40mg od po

35) **D**

The standard adult dose in triple therapy is **20mg od po**.

36) Which one of the following abbreviations means *four times daily*?

A) qh

B) qds

C) tds

D) nocte

E) od

36) **B**

The abbreviation, qds, stands for quater die sumendum. The literal translation of this latin phrase is four ingestions daily, i.e. **four times daily**.

**Table 1**

| Abbreviation | Latin | English |
|---|---|---|
| qh | quaque hora | Every hour |
| tds | ter die sumendum | Three times daily |
| --- | nocte | At night |
| od | omne in die | once a day |

(Continued in *Appendix 1*).

37) You are called to the bedside of 76 year old female patient who is a longstanding patient on a rehabilitation ward. The attending nurse is concerned by the patient's recent blood results; [K$^+$] = 7.0mM. Nothing in the patient's history accounts for this result. You elect to treat the patient by giving (i) 10ml of 10% calcium gluconate iv (ii) 50ml 50% glucose iv with 15 units of insulin iv (iii) 30g calcium resonium per

rectum (pr). Which of the following most accurately describes the function of calcium gluconate in this scenario?

(A) It stabilizes the electrical excitability of the heart.

(B) It stabilizes the electrical excitability of smooth muscle.

(C) It stabilizes the electrical excitability of nerve tissue.

(D) It stabilizes the electrical excitability of skeletal muscle.

(E) It has a placebo effect.

37) **A**

Calcium gluconate is given immediately to stabilize the **electrical excitability of the heart** and prevent sudden death by an arrhythmia.

38) In the above scenario what effect does giving insulin have on plasma [$K^+$]?

(A) It raises the plasma potassium concentration.

(B) It lowers the plasma potassium concentration.

(C) It has no effect on the plasma potassium concentration.

(D) It causes haemolysis of red blood cells to raise the plasma potassium concentration.

(E) It transiently raises the plasma potassium concentration and then returns it to physiological normality.

38) **B**

Insulin stimulates the sodium-potassium pump (i.e. the Na/K-ATPase) in the cell membrane to cause transport of potassium into the cell and sodium out of the cell. The extracellular fluid will equilibrate with the plasma and act to **lower the plasma potassium concentration**. When taking blood through a needle that is too small or a plunger that is moved too rapidly, haemolysis can occur and release intracellular potassium from the red blood cells. Consequently the haematology results will show an artefactually raised serum potassium concentration.

39) In the above scenario what is the function of the calcium resonium?

(A) It binds calcium ions and exchanges them with chloride ions.

(B) It binds calcium ions and exchanges them with potassium ions.

(C) It binds potassium ions and allows them to be absorbed by erythrocytes.

(D) It binds potassium ions and allows them to be expelled from the body.

(E) It binds calcium ions and exchanges them with sodium ions.

39) **D**

Calcium resonium (Calcium Polystyrene Sulfonate) **binds potassium ions** using non-covalent forces and acts as a sink for potassium ions that can then be **expelled from the body**. Calcium resonium is able to carry out this function because it is a chelator that is not absorbed by the human body.

40) As a result of your intervention in the above scenario, the patient survives. On the next ward round your consultant congratulates you. Later that day her registrar asks you to check the patient's medication to ensure that nothing that is being taken raises potassium levels. Which of the following drugs should you stop?

(A) Prochlorperazine

(B) Methotrexate

(C) Prednisolone

(D) Amiloride

(E) Paracetamol

40) **D**

**Amiloride** blocks the epithelial sodium channel (ENaC) in the distal part of the nephron; so sodium is not as readily reabsorbed from the epithelial side of the duct cells, which means that it is also less readily absorbed from the luminal side. Therefore more sodium is excreted by the kidneys, with water passively following the sodium. Amiloride does not cause potassium depletion; it does not cause potassium excretion from the kidneys. Hence amiloride is a potassium sparing diuretic. The continued use of this diuretic can lead to a hyperkalaemia.

41) Which of the following drugs has a well established acute side-effect of producing arrhythmias?
(A) Pravastatin
(B) Paracetamol
(C) Lignocaine
(D) Gold
(E) Penicillin

41) **C**

Any antiarrhythmic drug, if used above its therapeutic range, can cause arrhythmias. **Lignocaine** is an antiarrhythmic drug as well as a local anaesthetic. The Vaughan Williams classification designates lignocaine as a type 1 antiarrhythmic because of its inhibition of sodium channels in the myocyte cell membrane.

42) Co-trimoxazole is used to treat:
(A) Pneumocystis jiroveci infections
(B) Hayfever
(C) Gout
(D) Oral candida albicans infections
(E) Cushing's disease

42) **A**

Co-trimoxazole is a mixture of trimethoprim and sulphamethoxazole. Its classic use is in the treatment or prophylaxis of lung infections caused by **pneumocystis jiroveci** (pneumocystis carinii) invariably as a result of HIV infection and immunosuppression.

43) Pyridostigmine is useful in the treatment of myasthenia gravis because it:
(A) Attacks the antibodies that cause the disease.
(B) It binds to acetylcholine.
(C) It causes acetylcholinesterase to be endocytosed.
(D) It inhibits acetylcholinesterase.
(E) It stimulates the sympathetic nervous system.

## 43) **D**

Pyridostigmine binds to and **inhibits acetylcholinesterase** at the neuromuscular junction to prevent acetylcholine breakdown and increase the acetylcholine concentration at the nicotinic receptors; the increased acetylcholine level then favourably competes with the antibodies to the nicotinic receptors to facilitate neuromuscular transmission.

Nerve gases (e.g. sarin gas) are synthetic organophosporus compounds that are acetylcholinesterase inhibitors; they are lethal toxins at low doses. Because of their strong binding they are essentially irreversible inhibitors of the enzyme.

> **For this type of question, or any question related to neuromuscular junctions at skeletal muscle, remember that there are** nicotinic **receptors present at these junctions. They receive the chemical signal from acetylcholine at the postsynaptic membrane. It is these receptors that are responsible for neuromuscular transmission. Muscarinic receptors are not involved neuromuscular transmission** *at skeletal muscle.*

44) You are called to the bedside of a patient with a history of episodic atrial fibrillation. The patient's pulse is 155bpm. The patient's level of consciousness is currently reduced and has been varying for the last 48 hours in concert with the episodes of atrial fibrillation. You decide to use intravenous amiodarone to attempt to return the patient to sinus rhythm ("chemical cardioversion"). What is the clinically important molecular target of amiodarone?

(A) Bicarbonate channels

(B) Chloride channels

(C) Potassium channels

(D) Sodium channels

(E) Calcium channels

44) **C**

Amiodarone blocks **potassium channels** in the myocytes thus slowing the repolarisation of action potentials. This increases the duration of the depolarisation and repolarisation cycle as well as decreasing the maximum possible frequency of action potentials. Hence it has an antiarrhythmic effect - specifically it is a class III Vaughan Williams drug. Although there is pharmacological evidence that amiodarone also blocks sodium channels, it is believed that the dominant clinical effect is the action at potassium channels.

45) A patient is suspected to have taken an overdose of warfarin and laboratory results confirm a very high INR. Which of the following would be least helpful in reversing the effects of the warfarin?

(A) Vitamin K

(B) Fresh frozen plasma and intravenous vitamin K

(C) Fresh frozen plasma

(D) Prothrombin complex concentrate and aspirin

(E) Prothrombin complex concentrate

45) **D**

The **prothrombin complex concentrate** contains all of the factors that are sensitive to warfarin mediated inhibition of synthesis. The addition of **aspirin** to this mixture is not helpful. The aspirin will act as an anti-platelet aggregation factor and so increase the tendency to bleed. Vitamin K is used to counteract the effect of warfarin (as warfarin inhibits the synthesis of vitamin K dependent clotting factors). Fresh frozen plasma contains all of the clotting factors in normal plasma and so can be used therapeutically. The combination of fresh frozen plasma and intravenous vitamin K is also an effective clinical treatment.

| Vitamin K dependent Clotting Factors (vitamin K needed for their synthesis) |
|---|
| Factors X, IX, VII and II (*"1972"*) |

46) Whilst you are examining a teenager with appendicitis, the patient in the adjacent bed starts fitting. The attending nurse tells you that the patient was admitted urgently two days earlier because of a head injury and is awaiting results of a CT scan. You elect to deal with the pressing issue of the fit by giving which of the following drugs?
(A) Morphine
(B) Droperidol
(C) Haloperidol
(D) Diazepam
(E) Midazolam

46) **D**

The common anticonvulsant/antiepileptic drugs include phenytoin, valproate, carbamazepine and **diazepam**. Of these, only the benzodiazepine, diazepam, is represented in the above answer options. Midazolam is a short acting drug that is not generally used as an anticonvulsant/antiepileptic.

47) You have a cousin who has always suffered from excessive sweating and seeks your advice in managing the problem. He has tried normal antiperspirants but they were ineffective. Which of the following is most likely to be helpful?
(A) Aluminium chloride
(B) Botulinum toxin
(C) Wearing fewer clothes
(D) Noradrenaline
(E) Neostigmine

47) **B**

Aluminium chloride is present in normal antiperspirants and so is unlikely to be helpful here. However, **botulinum toxin** (produced by Clostridium Botulinum) is a neurotoxin that can be topically injected to prevent sweating. Botulinum toxin acts by binding presynaptically to high-affinity receptor sites on the cholinergic nerve terminals and decreasing the release of acetylcholine at the synapse. The effect can last to 5-10 months. It has an adverse effect of also inhibiting neuromuscular transmission to cause muscle weakness. It may not always be practical to wear fewer clothes (and the option implies that the individual is wearing clothes to remove). Beta blockers can be effective in diminishing the sympathetic drive to sweat but noradrenaline would exacerbate the problem. Similarly, neostigmine would increase the cholinergic transmission at the synapse by inhibiting acetylcholinesterase, thus also exacerbating the problem of excessive sweating.

48) A patient who has just started chemotherapy is feeling nauseous. She asks for something to prevent her from feeling "sick" or vomiting. Which of the following is most likely to be helpful?

A) Domperidone 12.5mg im

B) Cyclizine 50mg tds iv/im or po

C) Aluminium hydroxide

D) Haloperidol 3mg im

E) Drinking a lot of water.

48) **A**

The nausea and vomiting may be induced by chemotherapy, radiotherapy or the malignancy itself. **Domperidone** is a dopamine antagonist that is an established antiemetic used in chemotherapy. Cyclizine is an antihistamine that is used as an antiemetic adjunct to opiates, but is not particularly effective in chemotherapy. Aluminum hydroxide is an antacid used to manage excessive acidic secretions in the stomach. The haloperidol is likely to sedate the patient rather than act predominantly as an antiemetic.

49) Which of the following pairings of drugs and mechanisms of action, is incorrect?

(A) Lactulose – osmotically active laxative that increases water retention in the stool.

(B) Digoxin – antiarrhythmic that inhibits the Na$^+$/K$^+$-ATPase in the cell membrane.

(C) Chloramphenicol - binds to the 50S ribosomal subunit and inhibits the peptidyl-transferase.

(D) Metformin – increases insulin synthesis in the β cells of the Islets of Langerhans.

(E) Spironolactone – blocks the mineralocorticoid receptor, which would otherwise stimulate sodium re-uptake and potassium loss in the collecting duct.

49) **D**

Metformin acts at the liver to decrease glucose production and export by inhibiting gluconeogenesis. It does **not** increase insulin synthesis by the pancreatic β **cells of the Islets of Langerhans**. The other pairings are correct.

50) A 36 year old intravenous drug user spikes a fever of 38.5°C and on the same day develops a new murmur. He has his own native heart valves (not synthetic or transplanted valves). You decide that this is an acute presentation of bacterial endocarditis and send off blood cultures. Before the cultures return you initiate treatment with flucloxacillin (2g 4-hourly iv) and gentamicin (80mg 8-hourly iv). How does the gentamicin work?

(A) It binds to mRNA to prevent protein synthesis.

(B) It binds to the 50S ribosomal subunit in the affected bacteria and causes misreading of mRNA and aberrant initiation of protein synthesis.

(C) It binds to the 30S ribosomal subunit in the affected bacteria and causes misreading of mRNA and aberrant initiation of protein synthesis.

(D) It binds to tRNA to prevent protein synthesis.

(E) It inhibits transcription in the affected bacteria.

50) **C**

Gentamicin binds to the **30S ribosomal subunit in the affected bacteria and causes misreading of mRNA and aberrant initiation of protein synthesis**. In addition, it causes aberrant termination of protein synthesis. Gentamicin enters the Gram negative cell by an active transport system based on oxidative metabolism. Gentamicin is also effective against staphylococcus aureus.

Gentamicin is an aminoglycoside that is often used in combination with a penicillin to cause synergistic killing of infecting bacteria. The drugs are more effective together than the sum of their individual effects. This combination of antibiotics is effective against Gram positive infections – which is particularly appropriate because in this scenario a streptococcal infection is the most likely cause of the infective endocarditis. It should always be remembered that gentamicin is *nephrotoxic* and *ototoxic* and accordingly is usually only used in serious clinical scenarios.

51) You are called to see a patient on long term chlorpromazine. You are concerned that the patient may be suffering from the side-effects of this drug. Which of the following is an established side-effect of chlorpromazine?

(A) Bradycardia
(B) Diarrhoea
(C) Mania
(D) Urinary incontinence
(E) Tardive dyskinesia

51) **E**

Chlorpromazine is a traditional antipsychotic medication and so can cause the side-effects listed in the adjacent table. **Tardive dyskinesia** is a movement disorder that can occur after several months of treatment that produces jerky writhing movements. Tardive dyskinesia occurs in 25% of patients taking long term antipsychotic medication. Involuntary tongue protrusion is a classic example of tardive dyskinesia.

## Table 1: Side-effects of traditional antipsychotics

| Side-effect | Description |
| --- | --- |
| **Acute dystonia** (extrapyramidal effect = $D_2$ receptor effect) | Abnormal postures and muscle spasms. |
| **Tardive dyskinesia** (extrapyramidal effect = $D_2$ receptor effect) | Jerky or writhing movements. |
| **Akathisia** (extrapyramidal effect = $D_2$ receptor effect) | Feeling of restlessness and compulsion to move. |
| **Neuroleptic malignant syndrome** | Fever, delirium, fluctuating heart rate and blood pressure, muscular rigidity. |
| **Dry mouth** (antimuscarinic effect) | --------- |
| **Blurred vision** (antimuscarinic effect) | --------- |
| **Constipation** (antimuscarinic effect) | --------- |
| **Urinary retention** (antimuscarinic effect) | --------- |
| **Confusion** (antimuscarinic effect) | --------- |
| **Tachycardia** (antimuscarinic effect) | --------- |
| **Sedation** (antihistaminic effect) | --------- |
| **Postural hypotension** (anti-alpha adrenergic effect) | --------- |
| **Hypothermia** | --------- |
| **Weight gain** | --------- |
| **Depression** | --------- |

52) To which class of drug could hydrocortisone be most accurately assigned?
(A) Toxin
(B) Glucocorticoid
(C) Mineralocorticoid
(D) Corticosteroid
(E) Hormone

52) **B**

Hydrocortisone is a **glucocorticoid**. Glucocorticoids are produced by the adrenal cortex (zona fasciculata) and act at glucocorticoid receptors; the later have a major role in controlling glucose metabolism. Hence the name *glucocorticoid* is derived from (**gluco**se + **cort**ex + ster**oid**). However the drug is usually used clinically for its anti-inflammatory and immunosuppressant effects.

Corticosteroid is a general term for that covers both mineralo**cort**icoids and gluco**cort**icoids – both are produced in the adrenal **cort**ex.

However, clinically many people use the word corticosteroid to refer to just glucocorticoids.

*Cortisol and hydrocortisone are the same substance.*

53) You are called to the bedside of a patient who you notice has a round puffy face, abdominal striae and a buffalo hump. Which drug is he likely to have been taking for a significant period of time?
(A) Prednisolone
(B) Noradrenaline
(C) L-dopa
(D) Growth hormone
(E) Actrapid

53) **A**

**Prednisolone** is a synthetic glucocorticoid that can cause Cushingoid features when used for long periods:

**Table 2:**

| Cushingoid features; side-effects of chronic glucocorticoid use | |
| --- | --- |
| 1. Psychosis | 14. Protuberant abdomen |
| 2. Depression | 15. Testicular atrophy |
| 3. Acne | 16. Menstrual imbalance |
| 4. Round face ("Moon face") | 17. Proximal muscle wasting (thin arms and legs) |
| 5. Cataracts | 18. Osteoporosis |
| 6. Glaucoma | 19. Peripheral neuropathy |
| 7. Fat redistribution ("Buffalo hump") | 20. Immune suppression – infections / TB reactivation |
| 8. Peptic ulcers | 21. Hyperglycaemia / Diabetes mellitus |
| 9. Abdominal striae | 22. Hypertension |
| 10. Thin skin | |
| 11. Easy bruising | |
| 12. Impaired wound healing | |
| 13. Sodium and water retention | |

54) What is the best description of cocaine's mechanism of action?

(A) Potentiates dopaminergic, noradrenergic and serotoninergic transmission in the CNS by inhibiting their re-uptake from the synaptic spaces.

(B) Inhibits dopaminergic, noradrenergic and serotoninergic transmission in the CNS by stimulating their enzymic degradation.

(C) Potentiates GABA transmission in the CNS.

(D) Inhibits cholinergic transmission in the CNS.

(E) Potentiates L-dopa transmission in the basal ganglia.

54) **A**

Cocaine **potentiates dopaminergic, noradrenergic and serotoninergic**

**transmission in the CNS by inhibiting their re-uptake from the synaptic spaces.**

Cocaine is a stimulating alkaloid obtained from the leaves of the coca plant that causes euphoria as well as tending to lead to addiction. Cocaine is also a local anaesthetic.

55) Which of the following is neither a natural nor a synthetic catecholamine?

(A) Isoproterenol
(B) Noradrenaline
(C) Adrenaline
(D) Dopamine
(E) Acetylcholine

55) **E**

**Acetylcholine** is a quaternary amine that does not have the chemical structure of a catecholamine. The catecholamines are so named because they contain a catechol or 1,2-dihydroxybenzene group, $C_6H_4(OH)_2$. The catecholamines dopamine, noradrenaline and adrenaline are the key neurotransmitters in the sympathetic nervous system. Isoproterenol (isoprenaline) is a synthetic catecholamine that is a selective agonist for $\beta_1$ and $\beta_2$ adrenergic receptors and so has been used as an antiarrhythmic and bronchodilator.

56) A patient is required to take digoxin regularly to manage persistent atrial fibrillation with a fast ventricular rate. Although the drug is being used successfully the patient wants to know why so many blood tests are required. Your answer should refer to:

(A) The narrow therapeutic window of digoxin.
(B) Monitoring for an allergic response.
(C) Evidence of rhabdomyolysis.
(D) Measuring albumin levels.
(E) Measuring free $Fe^{2+}$ levels.

56) **A**

Digoxin has a very **narrow therapeutic window**; plasma concentrations that are too low may be ineffective and concentrations that are too high may be toxic. Target concentrations are usually 1.0-1.5 nmol/l. The most worrying toxic effects are cardiac arrhythmias.

57) Which of the following pairings of drugs and mechanisms of action, is incorrect?

(A) Bupivacaine; inhibits voltage sensitive sodium channels.

(B) Suxamethonium; causes muscle paralysis by acting as an agonist at nicotinic receptors in the neuromuscular junction.

(C) Ecstasy (MDMA); the euphoria is caused by inhibition of serotonin re-uptake in the CNS and increased serotonin levels in the synaptic cleft.

(D) Diclofenac; a specific COX-1 inhibitor.

(E) Pravastatin; inhibits cholesterol synthesis and decreasing circulating LDL levels.

57) **D**

Diclofenac is **not a specific COX-1 (cyclooxygenase 1) inhibitor**. It inhibits COX-1 and COX-2 and so can increase the risk of peptic ulceration. Cyclooxygenase 1 is important for normal function of gastric mucosa. As the ending –caine suggests, bupivacaine is a local anaesthetic and this class of drug prevents action potentials from propagating by inhibiting the voltage gated sodium channel in the cell membrane. Suxamethonium is a skeletal muscle relaxant used as an adjunct to general anaesthesia – it acts as an agonist. However, both agonists and antagonists at the nicotinic receptor can cause the necessary paralysis.

Ecstasy (MDMA; 3,4-methylenedioxymethamphetamine) has a euphoric effect that is caused by the inhibition of serotonin re-uptake in the CNS synapses and so increases serotonin levels in the synaptic cleft. Ecstasy's other effects are mediated by potentiating noradrenaline and dopamine neurotransmission in the CNS. Pravastin's metabolic effect in inhibiting HMG-CoA reductase means that it slows the rate determining step in cholesterol synthesis. The reduced levels of cholesterol in the liver cause more absorption from the circulatory system – thereby diminishing plasma LDL levels.

58) An adult patient has cystic fibrosis and takes enzyme supplements to maintain near normal digestion. Which of the following enzymes would you most expect to be in the supplement?

(A) Amylase
(B) Pepsin
(C) Lipase
(D) Protein kinase A
(E) Protein kinase C

58) **C**

Cystic fibrosis sufferers can still produce salivary amylase and gastric pepsinogen. However the failure of secretion of pancreatic enzymes means that little **lipase** is available for fat digestion. (In adults, gastric lipase probably accounts for less than 20% of total lipase activity. In addition this lipase can only break down triglycerides to diglycerides – the latter cannot be absorbed). Thus the most important digestive enzyme to include in the supplement is lipase. Protein kinase A and C are not digestive enzymes but are important in intracellular signal transduction.

59) An overweight 50 year old woman has undergone a recent total abdominal hysterectomy with bilateral salpingo-oophorectomy (TAH and BSO). Three days after the procedure she complains about pain and swelling in her left calf. You confirm the localized swelling and a D-dimer test is positive. Compression stockings are applied to the left leg. You start her on heparin and warfarin anticoagulation. What is the mechanism of action of heparin?

(A) Heparin binds to antithrombin III and increases its activity.
(B) Heparin binds to antithrombin III and decreases its activity.

(C) Heparin prevents the synthesis of vitamin K dependent clotting factors.

(D) Heparin increases the synthesis of vitamin K dependent clotting factors.

(E) Heparin hydrolyses fibrin.

59) **A**

Heparin's primary mechanism of action requires **binding to antithrombin III to increase its activity**. Heparin causes a conformational change in antithrombin III that increases its binding to the active forms of key clotting factors i.e. 12a, 11a, 10a, 9a and 7a. Antithrombin III inhibits their actions by binding irreversibly to these factors and so prevents the action of the clotting cascade. Option (C) is a description of warfarin's mechanism of action. Warfarin works through inhibition of the biosynthetic enzyme, vitamin K epoxide reductase. Because warfarin works at the level of protein synthesis, whereas heparin acts on pre-formed proteins, warfarin takes longer to act and has a more prolonged effect. Hence when dealing with a DVT, heparin is the first drug to become effective, despite being given at the same time as the warfarin.

60) Why is heparin only given as an initial treatment?

(A) Because it is faster acting than warfarin and is usually given as an injection or as a drip whilst the patient is in hospital.

(B) Because heparin is cheaper than warfarin.

(C) Because heparin has a low bioavailability.

(D) Because heparin is hepatoxic after three weeks.

(E) Because heparin is cardiotoxic after two weeks.

60) **A**

**Heparin is faster acting than warfarin and is usually given as an injection or as a drip** – meaning that heparin is easier to manage as an *inpatient*. Obviously, the patient is an inpatient at the beginning of the development of the DVT and outpatient later on. Heparin is more expensive to administer because it is usually given on an inpatient basis. Heparin has a high bioavailability because it can be given intravenously. Heparin is not recognized as being significantly hepatotoxic or cardiotoxic.

# Short Answer Questions: Answers

## Question 3

A) As a junior doctor working for a surgical firm, you have to manage many postoperative patients. It is not unusual for one of your patients to be pyrexial.

Name four causes (diagnoses) of postoperative fever. (4 marks)

**Table 1: Causes of Postoperative Fever – 5Ws**

| Water | Wind | Wound | Walking | Wonderdrug |
|---|---|---|---|---|
| Urinary Tract Infection | Pneumonia | Wound infection | Deep venous thrombosis and pulmonary embolism | Penicillin (many other drugs as well) |

B) What is the first line drug used in the treatment of a urinary tract infection (UTI)? (1 mark)

## Trimethoprim

C) What are the first line drugs used in the treatment of deep venous thrombosis (DVT)? (2 marks)

Low molecular weight **heparin**
**Warfarin**

D) Name the tests used to show that heparin is within its therapeutic range and that warfarin is within its therapeutic range. (2 marks)

Warfarin => Measure **INR** (International Normalized Ratio)
Heparin  => Measure **aPTT** (activated partial thromboplastin time)

E) A 55 year old patient has undergone a heart valve replacement with an artificial valve. As a result she is required to maintain lifelong anticoagulation with warfarin. The aetiology of her heart valve dysfunction is believed to be childhood rheumatic fever. For the first year after the operation the warfarin mediated control of INR is routine. However, after the second year progressively larger doses of warfarin are required to maintain the target INR. Interestingly, after the second year the patient's renal function and cardiac function remain good, however the hepatic function deteriorates insidiously. The patient reports that she was divorced by her husband shortly after the valve transplant and is currently dating again. When she puts down her handbag the physician hears the clinking of glass bottles.

What is the most likely explanation for deterioration in the control of her INR? (3 marks)

A very common cause of liver impairment in Europe and North America is **alcohol abuse**. Alcohol induces cytochrome P450 enzymes in the liver that can lead to increased clearance of medicines such as benzodiazepines, phenytoin and **warfarin**. Larger doses will thus be required to maintain the target INR.

F) If the warfarin levels are maintained at too high a level, to what types of complications will the patient be prone? (1 mark)

**Haemorrhagic** disease (e.g. bruising/haemorrhagic stroke/haemarthrosis).

G) Over the course of the next five years, she develops spider naevi, caput medusae and haemorrhoids. Blood tests show markedly raised AST, ALT and ALP. Concurrently her personal and social life leave her unhappy and she admits that drinking excessive alcohol is a problem. What affective disorder can chronic alcohol abuse cause? (1 mark)

**Depression.**

H) The GP is concerned about the progression of her depression and decides that she needs support from antidepressants. He considers using an SSRI, SNRI and MAOI.
Give an example of each of these classes and briefly describe the primary mechanism of action. (6 marks)

**ssRI** - e.g. **Fluoxetine** – inhibits serotonin re-uptake in the CNS synapses.
**snRI** - e.g. **Venlafaxine** – inhibits serotonin and noradrenaline re-uptake in the CNS synapses.
**MAOI** – e.g. **Selegiline** – inhibits breakdown of catecholamines in the CNS.

## Question 4

A shot and wounded soldier is taken to an emergency medical facility. Because of her significant blood loss, it is decided to give her a transfusion. The soldier is unconscious and is unable to state her blood type, nor is she carrying anything that mentions her blood type.

A) She needs at least three units of blood. What blood type would you give her? (2 marks)

**O-** (2 marks)
Only 1 mark for "**O**"

B) In the makeshift hospital there is a chart that matches recipients to acceptable donors. However, not all of the boxes in the chart can be read. Complete the empty boxes:

| Red blood cell compatibility table | | | | | | | | |
|---|---|---|---|---|---|---|---|---|
| Recipient | Donor | | | | | | | |
| | O- | O+ | A- | A+ | B- | B+ | AB- | AB+ |
| O- | ✓ | | | | ✗ | ✗ | ✗ | ✗ |
| O+ | ✓ | | ✗ | ✗ | ✗ | ✗ | ✗ | ✗ |
| A- | ✓ | | ✓ | ✗ | ✗ | ✗ | ✗ | ✗ |
| A+ | ✓ | | ✓ | ✓ | ✗ | ✗ | ✗ | ✗ |
| B- | ✓ | | | | | | | |
| B+ | ✓ | | ✗ | ✗ | ✓ | ✓ | ✗ | ✗ |
| AB- | ✓ | | ✓ | ✗ | ✓ | ✗ | ✓ | ✗ |
| AB+ | ✓ | | ✓ | ✓ | ✓ | ✓ | ✓ | ✓ |

(8 marks)

| Red blood cell compatibility table | | | | | | | | |
|---|---|---|---|---|---|---|---|---|
| Recipient | Donor | | | | | | | |
| | O- | O+ | A- | A+ | B- | B+ | AB- | AB+ |
| O- | ✓ | ✗ | ✗ | ✗ | ✗ | ✗ | ✗ | ✗ |
| O+ | ✓ | ✓ | ✗ | ✗ | ✗ | ✗ | ✗ | ✗ |
| A- | ✓ | ✗ | ✓ | ✗ | ✗ | ✗ | ✗ | ✗ |
| A+ | ✓ | ✓ | ✓ | ✓ | ✗ | ✗ | ✗ | ✗ |
| B- | ✓ | ✗ | ✗ | ✗ | ✓ | ✗ | ✗ | ✗ |
| B+ | ✓ | ✓ | ✗ | ✗ | ✓ | ✓ | ✗ | ✗ |
| AB- | ✓ | ✗ | ✓ | ✗ | ✓ | ✗ | ✓ | ✗ |
| AB+ | ✓ | ✓ | ✓ | ✓ | ✓ | ✓ | ✓ | ✓ |

- **AB** individuals have both A and B antigens on the surface of their erythrocytes, however the blood plasma does not contain any antibodies against either A or B antigens. So an individual with type AB+ blood can receive blood from any group, but can donate blood only to another type AB individual. *AB+ individuals are universal recipients.*

- **A** individuals have the A antigen on the surface of their erythrocytes with blood serum containing antibodies against the B antigen. So a group A individual can receive blood only from individuals of types A or O and can donate blood to individuals with type A or AB.

- **B** individuals have the B antigen on the surface of their erythrocytes with blood serum containing antibodies against the A antigen. So a group B individual can only receive blood from individuals of group B or O and can donate blood to individuals with type B or AB.

- **O** individuals have neither A or B antigens on the surface of their erythrocytes, and their blood serum contains antibodies against A and B antigens. So a group O individual can receive blood only from a group O individual, but an O- person can donate blood to individuals of any ABO blood group (i.e., A, B, O or AB). *O- individuals are universal donors.*

C) Shortly after the patient is transfused she spikes a fever of 38.5°C and becomes tachycardic, tachypnoeic and hypotensive. You decide that the patient is suffering from an acute transfusion reaction. How is this most likely to have occurred? (3 marks)

The patient was accidentally transfused with the **wrong ABO blood group**, because of a **labelling error** or **a retrieval error**.

D) The transfusion is stopped, an iv saline infusion is commenced and furosemide given. An expert on blood transfusions is contacted. The subsequent disseminated intravascular coagulation (DIC) is managed by transfusing platelets. Complete the donor/recipient transfusion table below to minimize acute and chronic complications: (7 marks)

| Recipient | Donor | | | |
|-----------|-------|-----|-----|-----|
| | O | A | B | AB |
| O | ✓ | ✗ | ✗ | ✗ |
| A | ✗ | | ✗ | ✗ |
| B | | | ✓ | |
| AB | ✗ | | | |

Because platelets express ABO antigens, the **ideal transfusions are from the same ABO blood group**. Non-matched platelets can be given but there is a small chance of a transfusion reaction.

| Recipient | Donor | | | |
|-----------|-------|-----|-----|-----|
| | O | A | B | AB |
| O | ✓ | ✗ | ✗ | ✗ |
| A | ✗ | ✓ | ✗ | ✗ |
| B | ✗ | ✗ | ✓ | ✗ |
| AB | ✗ | ✗ | ✗ | ✓ |

# Viva Voce Questions
# Answers

14) Theoretically, what types of effect can a failing liver have on drug action?
a) An increased bioavailability because of a **lesser** *first pass effect,* so more drug enters the venous circulation.
b) A decreased concentration of serum proteins causes a decrease in drug binding to proteins, leading to **more free form of the drug in the circulation**.
c) A **decreased rate of catabolism** of the drug in the liver. This decreased clearance of the drug leads to higher active drug levels for longer periods of time.

15) Before a gastroscopy an anxious patient is given midazolam. What type of drug is midazolam and how does it act?
It is a short acting benzodiazepine which enhances binding of GABA to GABA$_A$ receptors. This magnifies GABA's inhibitory effect on neurons to cause sedation.

16) For which disease is the drug cocktail HAART used?
AIDS. (**H**ighly **A**ctive **A**nti-retroviral **T**herapy).

17) Define drug efficacy.
The capacity of a drug to activate a receptor once it is bound to the receptor.

18) Who is at most risk of diabetic ketoacidosis?
Type 1 diabetics.

19) Name a class of drug that will never start fires.
NSAIDS – non-steroidal *anti-inflammatory* drugs. ☺

20) Name four classes of drugs that can be used to manage hypertension.

Diuretics, β blockers, α blockers, calcium channel blockers, ACE inhibitors, angiotensin II receptor blockers and direct renin inhibitors.

21) Methotrexate is an anticancer drug which is an anti-metabolite. What is an antimetabolite?

A drug that is an analogue for a natural metabolite but inhibits the normal metabolic process; antimetabolites are toxic. (In this case methotrexate inhibits *folate synthesis*).

22) Methotrexate can have a notorious adverse effect on the lungs. What is this effect?

Fibrosis.

23) Why do loop diuretics predispose to hypokalaemia?

Because by inhibiting the $Na^+/Cl^-/K^+$ transporter they prevent reabsorption of $K^+$ ions.

24) Name two classic uses of loop diuretics.

The management of:

Acute pulmonary oedema, chronic heart failure, hypertension, liver failure (ascites) and/or renal failure.

25) Which important sympathetic function is controlled by postganglionic muscarinic receptors?

Sweating.

26) Which drug is longer acting, salmeterol or salbutamol?

Salmeterol.

# Session 3

# Multiple Choice Questions
## Answers

| | | |
|---|---|---|
| 61) D | 71) A | 81) B |
| 62) D | 72) A | 82) C |
| 63) C | 73) A | 83) A |
| 64) D | 74) A | 84) E |
| 65) B | 75) A | 85) B |
| 66) C | 76) A | 86) E |
| 67) B | 77) D | 87) D |
| 68) E | 78) D | 88) C |
| 69) A | 79) A | 89) A |
| 70) C | 80) D | 90) D |

61) Intrathecal drug delivery implies the introduction of a needle into the subarachnoid space for the purpose of instilling a material for diffusion throughout the spinal fluid (CSF). Which of the following is a recognized clinical use of intrathecal drug delivery?
(A) Thrombolysis of CNS thrombi.
(B) Treating encephalitis.
(C) Treating meningitis.
(D) Control of chronic pain, both cancer and non-cancer related.
(E) Managing severe migraines.

61) **D**

Intrathecal drug administration is often used to manage intractable neuropathic pain or to palliate cancer pain. Intrathecal drug administration can cause meningitis but it is not used as a treatment for either meningitis or encephalitis. Intrathecal drug administration has yet to be used in the management of migraines or CNS thrombi.

62) What is the mechanism of action of heroin?

(A) Heroin acts indirectly at μ opiate receptors in the CNS to remove GABAnergic inhibition of noradrenergic neurotransmission.
(B) Heroin acts directly at μ opiate receptors in the CNS to remove GABAnergic inhibition of noradrenergic neurotransmission.
(C) Heroin acts directly at μ opiate receptors in the CNS to remove GABAnergic inhibition of dopaminergic neurotransmission.
(D) Heroin acts indirectly at μ opiate receptors in the CNS to remove GABAnergic inhibition of dopaminergic neurotransmission.
(E) Heroin potentiates serotoninergic neurotransmission.

62) **D**

Heroin and diamorphine are the same substance. When absorbed into the CNS (after crossing the blood-brain barrier), heroin is converted into morphine. Morphine is the form that acts at **μ opiate receptors in the CNS to remove GABAnergic inhibition of dopaminergic neurotransmission**, thus increasing the dopamine levels in the synaptic cleft and potentiating dopamine's action. Heroin's actions are particularly significant in the nucleus accumbens and the ventral tegmental areas of the brain – these areas form part of the brain's "reward pathway" and accounts for heroin's addictive features. The continued activation of the dopaminergic reward pathway accounts for the euphoria caused by the drug.

63) During a ward round, one of your asthmatic patients has a severe asthma attack. You initiate the protocol for managing such an attack and give (a) 60-100% oxygen, (b) 5mg salbutamol plus 500μg ipratropium, (c) 50mg prednisolone po and 100mg hydrocortisone iv. The patient recovers from the asthma attack. What is the purpose and mechanism of action of ipratropium?
(A) It is an anti-inflammatory that decreases the frequency of severe asthma attacks.
(B) It is a muscarinic that causes bronchodilation and prevents secretion from the bronchial epithelia.

(C) It is an antimuscarinic that causes bronchodilation and prevents secretion from the bronchial epithelia.

(D) It is a sympathomimetic that causes bronchodilation and prevents secretion from the bronchial epithelia.

(E) It is a β-adrenergic antagonist that causes bronchodilation and prevents secretion from the bronchial epithelia.

63) **C**

Ipratropium is an **antimuscarinic that causes bronchodilation and prevents secretion from the bronchial epithelia**.

64) Considering the above scenario, which adrenergic receptors are stimulated by salbutamol to cause the clinically useful effect?

(A) $\alpha_1$

(B) $\alpha_2$

(C) $\beta_1$

(D) $\beta_2$

(E) $\beta_3$

64) **D**

The acute bronchodilator effect is mediated through $\beta_2$ receptors in the bronchial tree. *In comparing this with the receptors involved in cardiac stimulation, it is helpful to remember that the reader has 1 heart ($\beta_1$) but 2 lungs ($\beta_2$).*

65) If, in the scenario above, the patient had not fully recovered after 15-30 minutes then 250mg aminophylline iv could have been administered. What is the function and mechanism of action of aminophylline?

(A) It is an anti-inflammatory that inhibits COX-1.

(B) It is a bronchodilator that acts through inhibition of phosphodiesterases.

(C) It is an anti-inflammatory that decreases bronchial secretions.

(D) It is a bronchodilator that acts through the stimulation of adenylate cyclase.

(E) It is a bronchodilator that acts through the stimulation of guanylate cyclase.

65) **B**

Aminophylline **is a bronchodilator that acts through inhibition of phosphodiesterases**. Aminophylline is a mixture of theophylline and ethylenediamine (the latter increases the water solubility of the active agent, theophylline). By inhibiting phosphodiesterases aminophylline slows the breakdown of secondary messengers such as cAMP and cGMP and potentiates their actions. As $\beta_2$ receptors cause bronchodilation by stimulating cAMP production it is understandable that inhibiting cAMP breakdown would also be an effective and synergistic means of causing bronchodilation. However, because aminophylline has a non-selective effect in increasing intracellular secondary messenger levels it is associated with many adverse effects. Aminophylline does not stimulate adenylate cyclase or guanylate cyclase. Aminophylline is not an anti-inflammatory.

66) A seven year old boy sees his GP several times in a year because of minor infections. He becomes noticeably more fatigued. On examination his liver and spleen are mildly enlarged and generalized lymphadenopathy is noted. Blood tests show anaemia. The GP is suspicious of acute lymphoblastic leukaemia (ALL). Blood smears and cytogenetics confirm the GP's suspicions. After referral to an oncologist the boy is started on a remission chemotherapeutic regime of vincristine, prednisolone, L-asparaginase and daunorubicin. What is the most likely complication of vincristine use?

(A) Diarrhoea

(B) Hypernatraemia

(C) Hair loss

(D) Blindness

(E) Central neuropathy

66) **C**

Vincristine inhibits mitosis through binding to tubulin to cause disruption of the formation of the cell cytoskeleton and mitotic spindle. So as well as affecting malignant cells, it will also affect rapidly turning over non-malignant cells. Hence **hair loss** (acting at hair follicles) and constipation (acting at intestinal epithelium) are common complications. **Hyponatraemia** and **peripheral** neuropathy are established adverse effects of vincristine use.

67) Why are four drugs used in the remission phase of this chemotherapy?

(A) Because four drugs are used in the maintenance phase of treatment.

(B) To increase the chances of rapidly killing at least 95% of the malignant cells.

(C) Using five drugs would be prohibitively expensive.

(D) To increase the chances of the development of drug resistant strains of the mycobacteria.

(E) To prepare for CNS prophylaxis by radiation therapy.

67) **B**

The purpose of the remission phase is to rapidly kill the vast majority of the malignant cells. In practice, this means killing about **95% of the malignant cells**. In recent years CNS prophylaxis by radiation therapy has fallen out of favour.

68) What is the mechanism of action of L-asparaginase?

(A) It increases the basal metabolic rate of the tumour cells.

(B) It facilitates the synthesis of the toxic substance asparagine.

(C) It potentiates the action of corticosteroids.

(D) It potentiates the action of daunorubicin.

(E) It hydrolyses asparagine that the leukaemic cells need for protein synthesis and growth and so is cytotoxic.

68) **E**

L-asparaginase **hydrolyses asparagine that the leukaemic cells need for protein synthesis and growth**. The drug is effective because leukaemic cells are unable to synthesize the amino acid asparagine, whereas normal cells are able to synthesize this amino acid. Hence the addition of L-asparaginase as a chemotherapeutic agent has a selectively negative effect on the malignant cell growth.

69) Which of the following drugs is not used in the initial treatment of TB infections?

(A) Erythromycin

(B) Isoniazid

(C) Ethambutol

(D) Pyrazinamide

(E) Rifampicin

69) **A**

The initial high intensity phase in the treatment of TB usually involves the four drugs:

Rifampicin, isoniazid, pyrazinamide and ethambutol *("RIPE")*.

Another first line drug that is sometimes used is streptomycin.

**Erythromycin** is not used in this anti-TB protocol.

70) Pethidine is a member of which of the following classes of drug?

(A) Anti-emetic

(B) Antidepressant

(C) Opioid analgesic

(D) NSAID analgesic

(E) Sedative

70) **C**

Pethidine is an **opioid analgesic**. Hence its dependence issues are similar to morphine.

71) A 45 year old man presents with severe dehydration and fatigue at your Accident and Emergency department. When you measure his blood glucose it is 35mM, there is a significant increase in ketones in his blood and the pH of the arterial blood is 7.25. You diagnose diabetic ketoacidosis (DKA) and initiate rehydration, potassium

supplementation and start intravenous insulin administration. Why is it necessary to give insulin?

(A) Classically DKA occurs in type 1 diabetics; these individuals cannot produce significant endogenous insulin.

(B) Individuals with DKA produce excessive amounts of proteases that degrade the insulin protein.

(C) Insulin is given to control the plasma potassium levels.

(D) Insulin is given to control the plasma calcium levels.

(E) Insulin is given to control the plasma sodium levels.

71) **A**

DKA most often occurs in **type 1 diabetics who cannot produce significant endogenous insulin** because of an autoimmune destruction of the beta cells of the Islets of Langerhans. Such individuals thus have difficulty transferring glucose from the plasma into the body's cells. The plasma glucose levels can rise to dangerously high levels causing osmotic diuresis and dehydration. Potential complications of rehydration are cerebral oedema and death.

*In the rare circumstances where DKA develops in type 2 diabetics, such individuals would also require insulin administration to control their blood glucose levels.*

72) You have admitted an adult patient with an erratic lifestyle who indulges in intravenous drug use and unprotected sex. On examination, anogenital ulcers are noted. Blood tests reveal treponeme-specific antibodies. You should initiate treatment with which of the following prescriptions?

(A) Doxycyline 200mg bd po for 28 days

(B) Cefuroxime 250mg bd po for 7 days.

(C) Co-amoxiclav 1.2g tds iv for 10 days.

(D) Metronidazole 2g od po for 10 days.

(E) Erythromycin 250mg qds po for 7 days.

72) **A**

The patient has primary syphilis (anogenital ulcers, positive antibodies and compatible lifestyle). The two possible standard first line treatments are benzylpenicillin or **doxycycline**.

73) A patient is on an oncology ward because of disseminated malignant melanoma. Overnight the patient complains of a new severe headache and associated nausea and vomiting. Ophthalmoscopy reveals enlarged retinal veins, loss of venous pulsation and haemorrhaging next to the optic disk. You give 150ml of 20% mannitol intravenously over 10 minutes. You increase the patient's diamorphine dosage and add 8mg dexamethasone at 12 hourly intervals, to be administered intravenously. Finally, you request a surgical consultation. What is the purpose of the mannitol?

(A) It is an osmotic diuretic that decreases extracellular fluid volume, so lowering intracranial pressure.

(B) It is a loop diuretic that decreases extracellular fluid volume, so lowering intracranial pressure.

(C) Mannitol is an effective analgesic with a short halflife.

(D) Mannitol is an anti-inflammatory medication.

(E) Mannitol is a cytotoxic chemotherapeutic agent.

73) **A**

This is an example of acutely raised intracranial pressure probably as a result of metastasis of the malignant melanoma to the brain. Mannitol is an **osmotic diuretic that generally decreases extracellular fluid volume, so drawing fluid from the intracranial space and hence lowering intracranial pressure**. This is possible because mannitol is a monosaccharide that forms hydrogen bonds with water, so mannitol draws a large volume of water with itself as it is excreted through the kidneys. Mannitol is not an analgesic, anti-inflammatory or chemotherapeutic agent.

74) What was the purpose of the dexamethasone in the previous scenario?

(A) It is a glucocorticoid that decreases the inflammation and hence inflammatory oedema, caused by the intracranial malignancy.

(B) It is a COX-1 inhibitor and an anti-inflammatory.

(C) It is an osmotic diuretic.

(D) It is a mineralocorticoid that decreases the inflammation and hence inflammatory oedema, caused by the intracranial malignancy.

(E) It is a radiotherapeutic agent.

74) **A**

Dexamethasone is a **glucocorticoid analogue that decreases inflammation and hence the inflammatory oedema, caused by the intracranial malignancy**. It is the most potent of the common glucocorticoid drugs. Dexamethasone binds to intracellular steroid receptors, diffuses to the nucleus and acts to control protein synthesis by acting as a transcriptional regulator. It acts to inhibit the production and action of a range of proteins/enzymes required for inflammation. This includes phospholipase A$_2$ required for the synthesis of arachidonic acid, which is a common precursor of several inflammatory mediators.

75) What is another name for diamorphine?

(A) Heroin

(B) LSD

(C) Amphetamine

(D) Methadone

(E) Ecstasy

75) **A**

**Heroin** is diamorphine. Methadone is the next most closely related drug because it is a synthetic opioid. Ecstasy, amphetamine and LSD are non-opioid recreational drugs.

76) A 16 year old boy who is a keen footballer presents to Accident and Emergency with a unilateral rash over the left shin. The rash is well defined, erythematous (red), painful and warm. There is no other

medical history of note. You make the diagnosis of cellulitis. Which is the most appropriate drug treatment?

(A) Flucloxacillin and benzylpenicillin

(B) Ceftazidime

(C) Ceftazidime plus metronidazole

(D) Trimethoprim

(E) Benzylpenicillin and metronidazole

76) **A**

Considering a patient who is not allergic to penicillin, the standard initial therapy for an uncomplicated cellulitis is a combination of **flucloxacillin and benzylpenicillin; together these are effective against staphylococcus aureus and streptococci**. Ceftazidime has only a limited activity against the most likely infective agent – staphylococcus aureus. Furthermore, cephalosporins are often used in the treatment of septicaemias and there is no evidence of a septicaemia in the question stem. Trimethoprim is commonly used to treat urinary tract infections (UTI) and there is no evidence of a UTI here; UTIs are usually caused by escherichia coli from the gastrointestinal tract and trimethoprim is effective against these organisms. Metronidazole is most effective against obligate anaerobes such as clostridium and bacteroides – it has little action against staphylococcus aureus.

77) Pilocarpine is a drug used to treat glaucoma. It acts through the stimulation of muscarinic receptors. Based on this information which of the following side-effects do you think pilocarpine would cause?

(A) Bronchodilation

(B) Pupillary dilation

(C) Decreased sweating

(D) Bradycardia

(E) Decreased salivation

77) **D**

Pilocarpine is a non-selective agonist at muscarinic receptors and so will cause parasympathomimetic complications. This question is actually asking the reader whether they are familiar with the actions of the parasympathetic nervous system. Pilocarpine causes **bradycardia** (it can be used to counter the effect of atropine of the heart). It also increases salivation, causes bronchospasm and constricts the pupils. Increased sweating is part of the function of the sympathetic nervous system, however the postganglionic neurons are stimulated by *muscarinic* receptors. So pilocarpine would, and does, increase sweating.

78) Which of the following is not a drug used to manage Parkinsonism?

(A) Bromocriptine

(B) Amantadine

(C) Selegiline

(D) Domperidone

(E) L-dopa

78) **D**

Individuals with Parkinsonism have impaired dopaminergic transmission in the CNS, particularly in the substantia nigra. Bromocriptine is a dopaminergic (D$_2$) agonist used to treat Parkinsonism. Amantadine stimulates dopamine release and is also used in the treatment of Parkinsonism. Selegiline is a MAO$_B$ inhibitor – this *mono*amine *o*xidase B inhibitor slows the breakdown of dopamine and so is also a useful treatment for Parkinsonism. L-dopa potentiates dopaminergic neurotransmission by being a precursor in the synthesis of dopamine.

**Domperidone** is an antagonist at D$_2$ receptors and so would exacerbate Parkinsonism – it is not used as a treatment.

79) A 65 year old man has a heart valve replacement that requires him to maintain appropriate anticoagulation levels, using warfarin as an anticoagulant. Over the last year his original warfarin dose has become less effective and this patient's INR (measure of anticoagulation) has begun to fall. You believe that the patient has been compliant with the

medication and is not taking any recreational drugs. There has been no significant change in his lifestyle – he continues to be an outgoing man with a lot of friends. His favourite activities are fishing and visiting the local pub. You decide to take a panel of baseline blood tests for renal and hepatic function. The tests show abnormal function of a major organ. What is the most likely cause of the decreased effectiveness of the warfarin?

(A) Chronic alcohol abuse has stimulated P450 in the liver to increase the clearance of warfarin.

(B) Chronic alcohol abuse has stimulated P450 in the liver to decrease the clearance of warfarin.

(C) Cirrhosis has caused an increase in the first pass effect.

(D) Impaired renal function has increased the warfarin excretion.

(E) Impaired liver function has decreased the warfarin excretion.

79) **A**

The patient is likely to be a chronic drinker of alcohol; alcohol increases the activity of the detoxifying **P450** enzymes of the liver. Warfarin is particularly susceptible to the change in activity of P450 and so is more readily broken down and cleared. Hence the warfarin becomes less effective. (The "abnormal function of a major organ" is likely to refer to the liver as the chronic alcohol use will predispose to steatosis, steatohepatitis and cirrhosis). Hepatic impairment would tend to decrease the first pass effect and increase drug availability – so the drug would apparently become *more* effective. Impaired renal function would not increase warfarin excretion (elderly people usually have decreased renal function and require less warfarin to achieve the same therapeutic effect).

80) Phase I drug reactions occur in the

(A) Heart

(B) Kidney

(C) Bladder

(D) Liver

(E) Brain

80) **D**

Phase I drug reactions are the first step in detoxification of a range of chemical moieties; this occurs in the **liver**. Phase II reactions involve conjugation in order to ensure water solubility of the drug and to facilitate excretion.

81) "The branch of pharmacology concerned with the movement of drugs within the body" is the definition of:

(A) Pharmacodynamics

(B) Pharmacokinetics

(C) Drug-receptor theory

(D) Pharmacosolution

(E) Pharmacodiffusion

81) **B**

This is the definition of **pharmacokinetics** (which can be alternatively stated as "the processes by which a drug is absorbed, distributed, metabolized and eliminated from the body"). Pharmacodynamics is the branch of pharmacology concerned with the biochemical and physiological effects of drugs and their mechanisms of action. Drug-receptor theory refers to agonist/antagonist and receptor interactions. The disciplines of pharmacosolution and pharmacodiffusion do not exist.

82) In pharmacology affinity refers to:

(A) The capacity of the drug to inactivate the receptor.

(B) The strength of the bond/attraction between the antagonist and receptor.

(C) The strength of the bond/attraction between the drug and receptor.

(D) The capacity of the drug to activate the receptor.

(E) The inverse of the speed of drug elimination.

82) **C**

Affinity is the measure of the **strength of the bond/attraction between the drug and receptor**. It can be expressed as an association constant, $K_A$ or as a dissociation constant, $K_D$.

83) Which of the following does not refer to drug receptor interactions?

(A) Compliance

(B) Agonist

(C) Affinity

(D) Antagonist

(E) Cooperativity

83) **A**

**Compliance** describes the degree to which a patient correctly follows medical advice – it does not relate to drug receptor interactions. An agonist binds to and stimulates its receptor whereas an antagonist binds to and inhibits its receptor. Affinity is the measure of the **strength of the bond (or attraction) between the drug and receptor**. It can be expressed as an association constant, $K_A$ or as a dissociation constant, $K_D$.

Cooperativity occurs if the binding of one molecule (or agonist/antagonist) to the receptor affects the ease of binding of another molecule/ion.

84) Which of the following is an SSRI?

(A) Senna

(B) Lithium

(C) Venlafaxine

(D) Selegiline

(E) Fluoxetine

84) **E**

**Fluoxetine** is an SSRI, a selective serotonin re-uptake inhibitor, commonly called Prozac. Lithium is commonly used in the treatment of manic-depressives (bipolar disease). Venlafaxine is a serotonin and noradrenaline re-uptake inhibitor (SNRI). Senna is a laxative.

85) Which is the most accurate primary mechanism of action of the combined oral contraceptive pill?

(A) Prevents implantation of the embryo.

(B) Suppresses release of gonadotrophins and ovulation.

(C) Prevents fertilization of the ovum by the sperm.

(D) Causes dyspareunia.

(E) Increases the frequency of miscarriage.

85) **B**

The combined oral contraceptive pill contains an oestrogen and a progestogen; these act to inhibit release of **gonadotrophins** (FSH and LH) and so prevent **ovulation**. The effect of the pill on implantation, fertilization or miscarriage is minor. The pill does not cause dyspareunia (painful intercourse).

86) Which of the following is not a side-effect of HRT therapy?

(A) Increased risk of thrombotic disease.

(B) Increased risk of uterine malignancy.

(C) Increased risk of a pulmonary embolus.

(D) Increased risk of a stroke.

(E) Increased risk of bladder malignancy.

86) **E**

Hormone replacement therapy, HRT, can increase the risk of:

- Primary breast malignancy.
- Primary endometrial malignancy.
- Primary ovarian malignancy.
- DVT and PE.
- Stroke.
- Ischaemic heart disease.
- Thrombotic diseases generally.

**HRT does not increase the risk of primary bladder malignancies**.

87) The proprietary name (brand name) for salbutamol is

(A) Rentopin

(B) Varent

(C) Salbutamone

(D) Ventolin

(E) Ergatol

87) **D**

Salbutamol was first sold under the brand name of **Ventolin**. Other names include **Ventoline, Ventilan, Aerolin** and **Ventorlin**.

88) Which disease/disorder is incorrectly matched with its drug management?

(A) Rheumatoid arthritis - Celecoxib

(B) Myocardial infarction - Aspirin

(C) Pulmonary embolus - Celecoxib

(D) Osteoarthritis - Paracetamol

(E) Febrile episode – Paracetamol

88) **C**

**Celecoxib** is a COX-2 selective NSAID often used in the management of osteoarthritis or rheumatoid arthritis. **Pulmonary embolus** is incorrectly matched with **celecoxib**; anticoagulation is achieved with warfarin and heparin. A COX-2 selective inhibitor is unlikely to be used as an acute analgesic in the management of pulmonary embolus.

89) Alcohol has which of the following effects when used *acutely*?

(A) Inhibits ADH release.

(B) Causes cirrhosis of the liver.

(C) Alcoholic cardiomyopathy.

(D) Haematemesis.

(E) Haemorrhage.

89) **A**

Alcohol has an acute effect of inhibiting ADH release through an action in the CNS – it thus behaves as a diuretic. (ADH = antidiuretic hormone = vasopression = arginine vasopressin). All of the other answer options are *chronic* effects of alcohol on the human body.

Haematemesis and haemorrhage are more likely to occur after severe chronic liver impairment.

90) You have a patient who has been smoking for 10 years and is finding it very difficult to stop. She asks you to explain the mechanism of action of cigarette smoke and why it is so difficult to quit. Which of the following statements is most important to include in your explanation?

(A) Nicotine from cigarettes makes breathing easier.

(B) Nicotine binds to nicotinic receptors in the brain to increase the levels of the neurotransmitter GABA in the pleasure centre of the brain.

(C) Nicotine can raise the metabolic rate and cause weight loss.

(D) Nicotine increases the level of the neurotransmitter dopamine in the pleasure centre of the brain.

(E) Nicotine can cross the blood brain barrier.

90) **D**

**Nicotine increases the level of the neurotransmitter dopamine in the pleasure centre of the brain**; the triggering of the reward centre (particularly the mesolimbic pathway) is responsible for the development of tolerance and ultimately addiction. The fact that nicotine can cross the blood brain barrier and that it can increase metabolic rate and weight loss are not directly relevant to the patient's issues of addiction. Nicotine's effects on ease of breathing are not significant acutely, however its chronic effects would make breathing more difficult (association with chronic obstructive pulmonary disease). Nicotine is not known to increase the levels of the neurotransmitter GABA in the pleasure centre of the brain.

# Short Answer Questions:
# Answers

## Question 5

A 58 year old businessman presents at Accident and Emergency complaining of upper abdominal pain. The pain has been present for 3 months and has worsened over the last week. He waited until after his merger before seeing a doctor; he has been taking Alka-Seltzer but this is becoming less effective. The pain was dull and is now sharp, is 7/10 in intensity and is associated with a disproportionate sense of fullness after a small meal. There is also associated nausea without vomiting. The businessman suffers from osteoarthritis of his left knee and regularly takes ibuprofen.

On examination epigastric tenderness is found. No other abnormalities are revealed. The patient is apyrexial with a regular pulse of 80 bpm and a blood pressure of 135/89.

A) What is the active component of the medication the businessman used to manage his stomach upset? (1 mark)

**Base** or **alkali** or **antacid** (1/2 mark)
**NaHCO₃** or **KHCO₃** (Sodium bicarbonate or potassium bicarbonate) (1 mark)

B) Considering *this* scenario, name three likely causes of the most probable gastric diagnosis. (3 marks)

**Stress**
**NSAIDs**
**Helicobacter pylori**
**Alcohol**

## Smoking

C) The patient is sent to endoscopy for examination of the stomach and lower oesophagus. Before the investigation, he is given midazolam. Where and how does this drug work? (3 marks)

Midazolam is a **short acting benzodiazepine** that acts predominantly in the central nervous system at **GABA$_A$ receptors** to potentiate the action of the **inhibitory neurotransmitter GABA** at these receptors. γ-Aminobutyric acid activity causes sedation, amnesia and muscle relaxation. The adverse effects predictably include confusion, coma, respiratory arrest and death.

D) The person administering the midazolam makes a dilution error and accidentally administers 100x the normal dose to this adult patient. Name 4 symptoms of midazolam overdose. (4 marks)

Any of:
Slow shallow breaths, cessation of breathing (*respiratory depression*).
Sleepiness, confusion, coma, death (*altered level of consciousness*).
Impaired coordination/balance, impaired reflexes, impaired speech/dysarthia (*impaired motor functions*).

Hypotension can also occur due to loss of vascular tone – although this is likely to be asymptomatic.

E) The medical team move rapidly to counteract the overdose by giving the patient intravenous flumazenil. Flumazenil acts by competitively inhibiting the binding of benzodiazepines at the GABA$_A$ receptor. The patient recovers and is well.
The endoscopic biopsies that were taken are sent to histopathology and

the report is eventually retrieved. It indicates that there is a malignancy underlying the gastritis. A surgical referral is made and the tumour excised and the patient is started on chemotherapy.

Name the two important characteristics of an idealized cocktail of chemotherapeutic drugs. (2 marks)

Perfectly **selective** for the malignant cells versus the host cells. (Hence few side-effects).
**100%** of the cancer cells are killed.

F) There are four subgroups of classical anticancer drugs that are distinguished by their mechanisms of action. Name the four subgroups of classical anticancer drugs. (4)

1 mark for each of the following:
**Alkylating agents**
**Antimetabolites**
**Cytotoxic antibiotics**
**Plant alkaloids**

G) The chemotherapeutic regime includes herceptin, 5-fluorouracil and doxorubicin. Briefly describe the mechanism of action of each drug. (3 marks)

- Doxorubicin – intercalates in DNA and hinders the action of topoisomerase and so *inhibits DNA replication*. (1 mark)
- Herceptin – is an antagonistic monoclonal antibody that binds to the overexpressed and overactive **Her2 protein** to switch off the mitogenic signal. (The erbB2 oncogene causes the synthesis of an aberrant EGF-like receptor that is permanently switched

on – this is the Her2 protein). (1 mark)
- 5-Fluorouracil is an antimetabolite that behaves as a toxic nucleotide analogue to prevent **DNA synthesis**. (1 mark)

## Question 6

A 47 year old amateur sportsman is admitted to his local Accident and Emergency department after complaining of chest pain. It is Sunday night and he spent the afternoon playing squash. The previous day he played football for three hours. The chest pain is described as 7/10 in intensity, central and radiates to his neck. It started two hours after he finished his game of squash, and is unusual in not being relieved by ibuprofen. The pain has continued and is still present on admission.

He has no family or personal history of arrhythmia, no evidence of hypertrophic obstructive cardiomyopathy (HOCUM) and no evidence of hyperlipidaemia. He has never smoked. He has no personal or family history of diabetes mellitus. His parents are in their 70s and are in good health – neither has suffered from angina or a myocardial infarction.

Palpation of the ribs or sternum could not elicit the pain.

The junior doctor managing the patient decides that the most likely diagnosis is a myocardial infarction.

A) Name four drugs routinely used to manage a myocardial infarction. (4 marks)

Oxygen, Nitrates (GTN), Clopidogrel, Aspirin, Morphine, Heparin (Enoxaparin), Streptokinase and Tissue Plasminogen Activator.

B) Clopidogrel and enoxaparin share a major side-effect (adverse effect). What is it? (1 mark)

An increased tendency to **haemorrhage**.

C) Describe the mechanism of action of enoxaparin. (3 marks)

Enoxaparin is a **low molecular weight heparin** that can be administered in a subcutaneous manner. Heparin's primary mechanism of action requires **binding to antithrombin III to increase the activity of antithrombin III**. Heparin causes a conformational change in antithrombin III that increases antithrombin III's ability to **bind to the active forms of key clotting factors i.e. 12a, 11a, 10a, 9a and 7a**. Antithrombin III inhibits their actions by binding irreversibly to them and so prevents the action of the clotting cascade.

The junior doctor requests an ECG and takes blood to test for the release of cardiac enzymes/troponin. The ECG changes are indeterminate with no clear ST elevation. However, troponin and CK-MB levels are significantly raised in the venous blood.

D) Why would a myocardial infarction raise circulating levels of troponin and CK-MB? (1 mark)

**The death of cardiac myocytes** causes a breakdown of the cell membrane and the release of intracellular contents into the adjacent tissue and blood. The troponin and cardiac enzymes are then released.

E) Although the patient finds the drug regime effective, he complains that one of the medications gives him a headache. Which medicine is it likely to be and why is the headache occurring? (2 marks)

**Glyceryl trinitrate**, GTN. (1 mark)

Headaches are believed to occur because GTN elevates the circulating **nitric oxide** concentration leading to smooth **muscle relaxation and vasodilation**. (0.5 mark each) GTN is both a venodilator and a vasodilator.

F) How and why are streptokinase and tissue plasminogen activator useful in the management of a myocardial infarction? (4 marks)

They are both **thrombolytic drugs** that breakdown the acutely occluding **blood clot** in the coronary artery.

Streptokinase – **facilitates the conversion of plasminogen to plasmin that lyses fibrin in blood clots.**
tPA – **catalyzes the conversion of plasminogen to plasmin.**

G) Three days after admission the patient is still suffering from a similar pain and has developed a pyrexia of 38°C. Angiography indicates that the three major coronary arteries are each more than 90% patent. The relatively young age and paucity of risk factors caused the consultant supervising the junior doctor to reconsider the working diagnosis. What is your new favoured diagnosis? (2 marks)

**Myocarditis** is the best diagnosis. (2 marks)
**Pericarditis** is less likely. (0.5 mark)
(Myocarditis and pericarditis can occur together).

H) What medications would you use to manage the latest diagnosis? (3 marks)

Reasonable combinations from:

**None** – most patients recover without drug intervention as this condition is usually self-limiting. (1 mark) or

**Paracetamol and NSAID** – Antipyrexial and analgesics. (2 marks) or

**Digoxin, Frusemide, Dobutamine, ACE inhibitor** – Heart failure medications. (3 marks)

Often the cause of the myocarditis is not found and most patients are managed supportively. Generally, physicians are understandably reluctant to take a cardiac biopsy from a patient suffering cardiac chest pain, even though this would lead to a definitive diagnosis. The causes, when are identified, are usually autoimmune or infective:

**Infections**

- Viral (e.g. Coxsackie virus, HIV, rubella virus, cytomegalovirus)
- Protozoan (e.g.Trypanosoma cruzi causing Chagas disease)
- Bacterial (e.g. Brucella)
- Fungal (e.g. Aspergillus)
- Parasitic (e.g. Schistosoma)

# Viva Voce Questions
# Answers

27) Name three glucocorticoid drugs.

Any three of the following commonly used drugs:

Dexamethasone, prednisolone, hydrocortisone, betamethasone, beclomethasone, fludrocortisone, methylprednisolone and deoxycorticosterone.

28) Can you name the class of drugs to which all of the following belong?
Nandrolone, testosterone, methyltestosterone, dihydrotestosterone and methandrostenolone.
Anabolic steroids = anabolic androgenic steroids

29) Which is longer acting, GTN or ISMN?
Isosorbide mononitrate is longer acting than glyceryl trinitrate.

30) Alendronate, risedronate and etidronate are members of which class of drugs?
Bisphosphonates. These are anti-osteoporosis medications.

31) What is miosis and what family of drugs is notorious for causing this?
Miosis is pupillary constriction. Miosis is the opposite of mydriasis. Opiates cause miosis.

32) Which enzymes are responsible for the initial breakdown of thyroid hormones?
Deiodinases.

33) For what is the drug simvastatin used?
It is used as a plasma cholesterol concentration lowering drug to decrease the risk of atherosclerotic disease.

34) What type of drug is actrapid?
A short acting insulin.

35) What is the common name for fluoxetine?
Prozac.

36) What effect does amiodarone have on thyroid hormones?
It can cause thyrotoxicosis by inhibiting deiodinases (the thyroid hormone breakdown enzymes). The high content of iodine in amiodarone can also lead to an increase in thyroid hormone synthesis (there are 75mg of iodine in a 200mg tablet of amiodarone).

37) Which of the following is not a medical emergency? Myocardial infarction, thyroid storm, malignant hypertension, hyperkalaemia and acute cranial arteritis.

*They are all medical emergencies.*

38) Name three negative issues that may complicate the use of NSAIDS.

Exacerbation of asthma, predisposition to cause renal failure and a predisposition to cause peptic ulceration.

39) What is another name for *neuroleptics*?

Antipsychotics.

40) A previously well 60 year old man develops a tremor, micrographia and clumsiness. Which drug would you test on the patient to confirm the suspected diagnosis?

A trial of Levodopa (L-dopa) could confirm the presence of Parkinsonism or Parkinson's disease.

# Session 4

# Multiple Choice Questions
## Answers

| | | |
|---|---|---|
| 91) A | 101) D | 111) C |
| 92) E | 102) B | 112) A |
| 93) B | 103) C | 113) E |
| 94) D | 104) C | 114) D |
| 95) C | 105) D | 115) A |
| 96) D | 106) E | 116) E |
| 97) B | 107) C | 117) B |
| 98) D | 108) A | 118) E |
| 99) C | 109) C | 119) D |
| 100) A | 110) E | 120) E |

91) Augmentin is

(A) Co-amoxiclav

(B) Cefuroxime

(C) Gentamicin

(D) Flucloxacillin and clavulanic acid

(E) Gentamicin and ampicillin

91) **A**

Augmentin is **co-amoxiclav** (amoxicillin and clavulanic acid); the clavulanic acid inhibits beta-lactamase activity. This reduces the risk of the occurrence of penicillin resistant bacteria.

92) Atropine is an antagonist at

(A) $M_1$ receptors

(B) $M_2$ receptors

(C) $M_3$ receptors

(D) $M_1 - M_3$ receptors

(E) All muscarinic receptors

92) **E**

Atropine is an antagonist at **all muscarinic receptors** (i.e. $M_1$-$M_5$).

93) Aniline dye workers have an increased the risk of

(A) Primary squamous cell carcinoma of the bladder.

(B) Primary transitional cell carcinoma of the bladder.

(C) Primary adenocarcinoma of the bladder.

(D) Small cell carcinoma of the bladder.

(E) Malignant melanoma of the bladder.

93) **B**

It is well established that **aniline dye** workers have an increased risk of developing **primary transitional cell carcinoma of the bladder**. It is believed that 2-naphthylamine, a derivative of aniline, increases the risk of occurrence of this bladder cancer.

94) Which of the following statements is true of oxytocin?

(A) Oxytocin secretion is decreased during labour.

(B) In late pregnancy the uterus becomes insensitive to oxytocin.

(C) Oxytocin inhibits the milk ejection reflex.

(D) Oxytocin can be given to initiate labour.

(E) Oxytocin has an anti-ADH effect at the kidney.

94) **D**

**Oxytocin can be given to initiate labour**. During pregnancy oestrogen increases uterine musculature sensitivity to oxytocin, which is accompanied by a marked increase in oxytocin receptors in the uterus. Oxytocin secretion is increased during pregnancy and the descent of the foetus into the birth canal causes a reflex release of additional oxytocin that facilitates parturition. Oxytocin is also important in initiating labour and can be used as a drug for that purpose. In addition, oxytocin has a physiological role in mediating the milk reflex – suckling at the breast causes release of oxytocin from the posterior pituitary. Finally, oxytocin and vasopressin (ADH) cross-react at their receptors, so oxytocin can have a significant ADH-like activity at the kidney.

95) A patient with symptomatic Wilson's disease and hepatic impairment is treated with penicillamine. The treatment is successful. Which of the following mechanisms accounts for the drug's success?
(A) Penicillamine acts as an antibiotic.
(B) Penicillamine binds to iron.
(C) Penicillamine binds to copper.
(D) Penicillamine binds to arsenic.
(E) Penicillamine is an antimetabolite.

95) **C**

Wilson's disease causes the accumulation of copper in the body, which results in hepatic impairment and neurological effects. Medical management makes use of a chelator (binder) of the **copper** that leads to its excretion from the body. Penicillamine is such a chelator. Penicillamine can also be used as a chelator in arsenic poisoning. Although penicillamine is a breakdown product of penicillin, it has no antibiotic activity. Penicillamine is not used as an iron chelator; deferoxamine or deferasirox can be used for iron. Penicillamine is a metabolite not an antimetabolite; classically, antimetabolites are used in chemotherapy as toxic analogues of normal physiologically important molecules.

| BINDER/Chelator | Use |
|---|---|
| Dimercaprol (BAL) | Acute arsenic poisoning<br>Acute mercury poisoning<br>Lead poisoning |
| Dimercaptosuccinic acid (DMSA) | Lead poisoning<br>Arsenic poisoning<br>Mercury poisoning |

| Dimercapto-propane sulfonate (DMPS) | Severe acute arsenic poisoning |
| | Severe acute mercury poisoning |
| Penicillamine | Copper toxicity |
| Ethylene diamine tetra-acetic acid (EDTA) | Lead poisoning |
| Deferoxamine and Deferasirox | Acute iron poisoning |
| | Iron overload |

96) A 33 year old man presents with malaise, generalized lymphadenopathy and tachypnoea. He is pyrexial with a temperature of 38°C. His chest X-ray image shows bilateral mid-zone shadowing. Acutely, he is treated with co-trimoxazole, which resolves the respiratory signs and symptoms. What was the acute disorder with which the patient presented?

(A) Streptococcal pneumonia

(B) Staphylococcal lobar pneumonia

(C) Bronchiolitis

(D) Pneumocystis pneumonia

(E) Jirovecal pneumonia

96) **D**

The patient presented with a **pneumocystis pneumonia** (caused by pneumocystis carinii/pneumocystis jiroveci). This is an opportunistic infection in the immunosuppressed. The mid-zone bilateral shadowing and response to co-trimoxazole treatment are characteristic of this infection.

97) Continuing the scenario above, the patient's underlying problems are managed by a cocktail of drugs including AZT. What is AZT?

(A) Adenozine triphosphate

(B) Zidovudine

(C) Azidothyroxine

(D) Zipovudine

(E) Adenine z-triphosphate

97) **B**

AZT is **zidovudine** (= azidothymidine). HIV is a retrovirus that carries its genomic material in the form of RNA, which is converted into DNA by its reverse transcriptase. The DNA is then inserted into the host's DNA for replication and subsequent viral assembly. AZT is a nucleoside analogue that binds to and inhibits the viral reverse transcriptase, preventing the viral DNA synthesis. None of the other answer options are drugs currently in existence.

98) Which of the following is not true of AZT?

(A) It is an antimetabolite.

(B) It is a reverse transcriptase inhibitor.

(C) It is a nucleoside analogue.

(D) It inhibits viral RNA synthesis.

(E) It inhibits viral DNA synthesis.

98) **D**

The HIV genome already exists as RNA; AZT inhibits viral DNA synthesis and does not **inhibit viral RNA synthesis**. Because AZT acts as a false nucleoside in the viral lifecycle it is effectively an antimetabolite when inhibiting the reverse transcriptase.

99) A 45 year old motorcyclist is rushed into A and E after crashing into a car at 50 mph. Signs and symptoms of internal bleeding are noted. As part of supportive management gelofusine is given before surgery. What type of substance is gelofusine?

(A) A laxative

(B) An osmotic diuretic

(C) A colloid

(D) An antibiotic

(E) A sedative

99) **C**

Gelofusine is a **colloid** that helps to maintain the circulatory fluid volume; it is often used under circumstances of acute and significant blood loss.

100) Which of the following is not an ACE inhibitor?

(A) Verapamil

(B) Captopril

(C) Capoten

(D) Altace

(E) Vasotec

100) **A**

**Verapamil** is a calcium antagonist not an ACE inhibitor. Captopril was the first ACE inhibitor. Capoten is a brand name for captopril. Altace is a brand name for the ACE inhibitor ramipril. Vasotec is a brand name for enalapril.

101) Which of the following is routinely used as a contrast agent in X-ray radiological investigations?

(A) Aluminium oxide

(B) Sodium nitrate

(C) Calcium sulphite

(D) Barium sulphate

(E) Magnesium sulphate

101) **D**

**Barium sulphate** is used to enhance contrast in X-ray based radiological investigations. For example, a "barium meal" can be used to aid examination of the upper gastrointestinal tract. This barium salt is relatively insoluble so little is absorbed by the human body – instead it is excreted by defaecation. Established adverse effects of the barium sulphate include nausea, vomiting, constipation and headaches.

102) Chlorhexidine is a widely used drug in hospitals and in homes. Most doctors use it at least once a day. What class of drug is

chlorhexidine?

(A) Diuretic

(B) Antiseptic

(C) Antispasmodic

(D) Antiarrhythmic

(E) Analgesic

102) **B**

Chlorhexidine is an **antiseptic** antibacterial with activity against Gram positive and Gram negative bacteria. It is present in skin cleaners, surgical scrubs, mouthwash (Corsodyl) and contact lens cleaner. The antibiotic action of chlorhexidine occurs through membrane disruption.

103) A 60 year old woman suffers from chronic joint pain. The digits of her hand show ulnar deviation. She is a smoker and her cells exhibit HLA DR4. Her daughter has recently developed joint pain in her hands as well. Which of the following drugs would you use to manage the mother?

(A) Silver

(B) Oestrogen

(C) Sulphasalazine

(D) Interferon

(E) Paracetamol

103) **C**

The patient has rheumatoid arthritis. **Sulphasalazine** is a disease-modifying drug (DMD) that can slow or stop the disease progression. Sulphasalazine and its metabolites (e.g. 5-aminosalicylic acid) are believed to be anti-inflammatory agents. However, sulphasalazine can take weeks to become effective and has the potentially serious side-effect of myelosuppression (bone marrow suppression). NSAIDS such as diclofenac and ibuprofen can be used as analgesics, but paracetamol is not usually effective.

104) At which receptor does actrapid exert its pharmacological effect?

(A) IGF-I receptor

(B) IGF-II receptor

(C) Insulin receptor

(D) EGF receptor

(E) PDGF receptor

104) **C**

Actrapid is an insulin preparation and exerts its key pharmacological effect at **insulin receptors**.

105) Which of the following is true about the mechanism of action of GTN?

(A) It causes vasoconstriction.

(B) It is converted to the active form, nitrous oxide.

(C) It causes dilation of veins only.

(D) It leads to the stimulation of guanylate cyclase.

(E) It decreases intracellular cGMP levels in vascular smooth muscle.

105) **D**

When absorbed GTN is converted into nitric oxide, NO (not *nitrous oxide*). Nitric oxide acts on receptors in vascular smooth muscle to **stimulate guanylate cyclase**. Guanylate cyclase synthesizes cGMP from GTP. The cGMP is a secondary messenger that activates the cGMP dependent protein kinase to initiate a phosphorylation cascade. This ultimately alters the phosphorylation profile of the contractile proteins to cause smooth muscle relaxation. GTN causes dilation of both veins and arteries.

106) Identify the established side-effect of ACE inhibitor use:

(A) Bradykinin depletion in the lungs.

(B) Hypokalaemia.

(C) Atrial fibrillation.

(D) Development of lymphoma.

(E) Precipitation or exacerbation of renal failure.

106) **E**

ACE inhibitors can cause **renal failure** by exacerbating the effects of renal artery stenosis. Under stable conditions the renal artery stenosis is balanced by action of the renin-angiotensin system to ensure sufficient afferent renal blood flow. The effect of the ACE inhibitor is ultimately to diminish this afferent blood flow and predispose to renal failure. ACE inhibition prevents the ACE mediated inactivation of bradykinin. The subsequent accumulation of bradykinin in the lungs can lead to a persistent cough. The diminished activity of aldosterone caused by ACE inhibitors leads to decreased potassium excretion and a tendency to hyperkalaemia. There is no association between ACE inhibitors and lymphomas or atrial fibrillation.

107) Which of the following drugs is an $\alpha$-adrenergic antagonist?

(A) Epinephrine

(B) Phenylephrine

(C) Prazosin

(D) Propranolol

(E) Atenolol

107) **C**

**Prazosin** is a selective $\alpha_1$-adrenergic antagonist that can be used to treat systemic hypertension. Phenylephrine is an $\alpha$-adrenergic agonist that is used as a nasal decongestant. Epinephrine is adrenaline, which is an $\alpha$-adrenergic agonist and a $\beta$-adrenergic agonist. Propranolol and atenolol are both non-selective beta blockers ($\beta$-adrenergic antagonists) and so act at both $\beta_1$ and $\beta_2$ receptors.

108) You find an individual collapsed and unconscious at the roadside. You arrive at the same time as the emergency medical personnel. The wife of the collapsed man tells you that he is prone to hypoglycaemia. You elect to manage the unconscious individual as if he were hypoglycaemic. Which of the following drug treatments would not be used?

(A) Oral glucose solution.

(B) Intravenous glucose solution.

(C) Intramuscular glucagon.

(D) Intravenous dextrose solution.

(E) Intravenous dextrose and saline.

108) **A**

An unconscious patient cannot be given **oral glucose solution**.

109) A 65 year old male ex-bricklayer presents at Accident and Emergency complaining of new shortness of breath. He is a longstanding smoker. On examination the patient is tachypnoeic and febrile. Prior to this episode he had a productive cough continuously for five months. He has no history of symptomatic cardiac disease. He has no other personal medical history of note. You request a chest X-ray and send off blood cultures as well as a taking blood for a full blood count, urea and electrolytes. At this point, what is the most likely working diagnosis?

(A) Pneumonia

(B) Chronic bronchitis

(C) Acute exacerbation of COPD

(D) COPD

(E) Acute exacerbation of asthma

109) **C**

The patient is a smoker who has had a productive cough for more than three months continuously (in two consecutive years) and so fulfils the clinical definition of chronic bronchitis. He has also presented with new onset pyrexia, dyspnoea and tachypnoea. This is consistent with an acute respiratory tract infection. Together these facts favour a diagnosis **of acute exacerbation of chronic obstructive pulmonary disease (COPD)**. This is a very common diagnosis in smokers. The diagnoses of pneumonia, COPD or chronic bronchitis alone are insufficient to account for all of the signs and symptoms. There is no evidence offered to support a definitive diagnosis of asthma.

110) Which of the following drugs would you not use to help the patient in the above acute scenario?

(A) Salmeterol

(B) Ipratropium bromide

(C) Oral prednisolone

(D) Amoxycillin

(E) Vancomycin

110) **E**

Acute exacerbation of COPD is an infective process that worsens background chronic lung disease. There is usually a small reversible component that will respond to bronchodilators such as the $\beta_2$ adrenergic agonist, salmeterol, or the anticholinergic, ipratropium bromide. Oral prednisolone is commenced as soon as possible to decrease the frequency of future episodes and to limit the duration of the current episode. The infective cause is usually bacterial so an antibiotic in the amoxycillin class is often given. The antibiotic may be changed in the light of the sensitivities shown by the blood cultures. **Vancomycin** is not a first line antibiotic and would not be used in the acute situation.

111) A 63 year old woman arrives at Accident and Emergency complaining of recent onset shortness of breath. There is no associated chest pain. She is an ex-smoker (20 pack years) with no other significant personal medical history. Her chest X-ray image shows a general fluffy infiltrate with prominent pulmonary vessels in the upper fields and peribronchial haze. Your impression is that this is pulmonary oedema. Which of the following is not a cause of pulmonary oedema?

(A) Vancomycin

(B) Levothyroxine

(C) Imiquimod

(D) Oestrogen

(E) Paracetamol

111) **C**

Renal failure, heart failure and liver failure are well known causes of pulmonary oedema. The release of pancreatic enzymes during acute pancreatitis can directly affect the lungs to cause inflammation and oedema. Oestrogen can cause acute pancreatitis by dramatically raising plasma triglycerides and so increasing the risk of biliary stones. A paracetamol overdose can cause acute liver failure through the accumulation of reactive toxic metabolites such as $N$-acetyl-$p$-benzoquinone imine that can cause hepatic necrosis. Vancomycin is notorious for having the potential for acute nephrotoxicity but the mechanism of action is unclear. Excessive levothyroxine results in thyrotoxicosis that can cause hyperdynamic heart failure (high output heart failure). Imiquimod is a topically applied immune stimulant that does not cause heart, liver or renal failure.

112) Considering the scenario above, how could frusemide (furosemide) help to manage the pulmonary oedema?

(A) Frusemide can reduce the fluid overload causing pulmonary oedema by increasing renal sodium and water excretion.

(B) It decreases the rate of coronary artery atherosclerosis.

(C) It acts on pulmonary arteries to cause vasoconstriction.

(D) It is a negative inotrope and chronotrope.

(E) Frusemide can reduce the fluid overload causing pulmonary oedema by increasing iron loss from the kidneys.

112) **A**

Frusemide is a loop diuretic that **can reduce the fluid overload causing pulmonary oedema by increasing renal sodium and water excretion.**

113) Assuming the pulmonary oedema originates in heart dysfunction, how might the β₁ agonist dobutamine, help in the management of pulmonary oedema?

(A) By stimulating dopamine receptors to increase noradrenaline release.

(B) By acting as a parasympathomimetic.

(C) By acting as a CNS neurotransmitter.

(D) By causing peripheral vasoconstriction to raise systemic blood pressure.

(E) By managing heart failure through positive inotropic and chronotropic effects.

113) **E**

Dobutamine is a selective $\beta_1$ agonist that improves cardiac function by positive **inotropic and chronotropic effects**. It does not function through or cause any of the other listed effects.

114) How does alendronate achieve its major therapeutic effect?

(A) It raises serum calcium ion concentrations.

(B) It stimulates vitamin D action.

(C) It increases calcium absorption from the GI tract.

(D) It inhibits osteoclast mediated bone resorption.

(E) It stimulates osteoblast action and bone deposition.

114) **D**

Alendronate **inhibits osteoclast mediated bone resorption**. With the inhibition of bone breakdown, bone synthesis becomes dominant. Subsequently bone density increases, and the likelihood of vertebral and forearm fractures decreases.

Bisphosphonates act as toxic metabolic analogues of pyrophosphate (they are effectively antimetabolites) to cause osteoclast death.

115) A previously well 25 year old medical student went on holiday to Ghana in West Africa. He was careful to take all of the appropriate vaccinations before departing to Africa. The student also took oral chloroquine commencing one week before departure and ceasing four weeks after he returned. Six weeks after returning from his holiday he attended his GP surgery complaining of a bad flu. On questioning by

his GP, the student admitted that he had an intermittent fever, which seemed to be worse every third day. On examination, hepatomegaly and splenomegaly were noted. No lymphadenopathy was found. The General Practitioner took blood samples that showed anaemia and hyperbilirubinaemia. Why did the medical student become unwell?
(A) Because of protozoal drug resistance.
(B) Because he failed to obtain the appropriate vaccination.
(C) Because he wore long-sleeved clothes.
(D) Because he had unsafe sex with a local prostitute.
(E) Because he drank dirty water.

115) **A**

The student has a febrile illness with a *three day period* after a visit to West Africa; this is malaria. Malaria is contracted after a bite from the female Anopheles mosquito that injects the plasmodium protozoa into the blood. **Protozoal drug resistance** to chloroquine is very common and has probably occurred here. A more effective prophylactic regime would have involved either mefloquine, doxycycline or malarone. There is no vaccination against malaria. Long sleeved clothes offer some protection against the mosquito bites. Malaria is not contracted through drinking dirty water or via sexual activity.

116) As a junior doctor working in Neurology you admit a patient with suspected herpes simplex encephalitis. Which of the following drugs would you give the patient in the acute scenario?
(A) HAART
(B) Gentamicin
(C) Ampicillin
(D) Prednisolone
(E) Acyclovir

116) **E**

**Acyclovir** (aciclovir) is the first line treatment for herpes simplex encephalitis. However, even with treatment there is a 20% mortality.

This antiviral drug is a guanosine nucleoside analogue toxic to the herpes virus. HAART is an anti-retroviral cocktail of drugs used to manage an HIV infection. Gentamicin and ampicillin are antibacterials. Prednisolone would exacerbate the herpes simplex infection because of its immunosuppressant action.

117) Which of the following analgesic regimes is most likely to be used in the management of mild pain?

(A) 10mg morphine prn

(B) 1g paracetamol qds po

(C) 30mg codeine qds po

(D) 1g paracetamol qds po and 75mg diclofenac bd po

(E) Co-codamol and diclofenac

117) **B**

Mild pain may be adequately managed by 1g paracetamol qds po. The analgesic ladder is described below:

| Severity of pain | Management |
|---|---|
| Mild | Paracetamol 1g 6-hourly and/or NSAID (e.g. diclofenac 75mg 12-hourly po) |
| Moderate | Add opioid to mild regime, e.g. codeine 30-60mg 6-hourly po. Continue paracetamol and/or NSAID. |
| Severe | Substitute a more potent opioid, e.g. morphine 5-10mg 4-hourly or prn. Continue paracetamol and/or NSAID. |

118) A 37 year old man presents at his General Practitioner's surgery complaining of a widespread red rash that is itchy, swollen and crusting. There is no obvious pattern to the distribution of the rash. There is no pyrexia or systemic symptomatology. He has no other significant past medical history. Despite an extensive history taking no

precipitating factor is identified. His mother has asthma and his father suffers from hayfever. You make a working diagnosis. Which of the following would not be part of the management?

(A) Discussion, education and reassurance.

(B) Topical hydrocortisone.

(C) Emollients (moisturizing cream).

(D) Calamine lotion.

(E) Ultraviolet light.

## 118) E

The description of the skin rash and the patient's family history are consistent with an *atopic eczema*. Atopic individuals are predisposed to asthma, hayfever, eczema and allergies. Atopy has a strong hereditary association. At a molecular and cellular level, such individuals are prone to an excessive IgE response to allergens. It is not clear why this happens but the hygiene hypothesis postulates that excessive cleanliness during childhood decreased the exposure to microbes necessary to stimulate and cause normal development of the immune system. Because eczema is an inflammatory process caused by an immune reaction, topical glucocorticoids can be effective in the management of the disease. A dry eczema can be managed with moisturizing creams. Calamine is an antipruritic (anti-itching) treatment. Patients benefit from discussion, education and reassurance regarding the disease. **Ultraviolet light** is not part of the management of eczema – it is used in the management of psoriasis.

119) Which of the following drugs can be used to most successfully treat an MRSA infection? (MRSA = Methicillin Resistant Staphylococcus Aureus)

A) Benzylpenicillin

B) Cephalexin

C) Amoxicillin

D) Vancomycin

E) Lancomycin

119) **D**

**Vancomycin** is used to treat MRSA infections. Beta-lactam antibiotics cannot be used to treat MRSA because the bacteria carry an enzyme (beta-lactamase) that can inactivate the antibiotic and allow the bacteria to thrive. Hence penicillins and cephalosporins are ineffective. Lancomycin does not exist.

120) Which of the following substances is not used to manage pre-eclampsia?

(A) Magnesium sulphate

(B) Nifedipine

(C) Labetalol

(D) Nicardipine

(E) Calcium sulphate

120) **E**

The key issues with pre-eclampsia are the rapid rise in blood pressure and the possibility of seizures. Magnesium sulphate is given to prevent fits; it is believed to cause CNS depression and to act as a calcium antagonist to cause vasodilation. Nicardipine and nifedipine are calcium antagonists used to manage the hypertension. Labetalol is a mixed $\alpha$ and $\beta$ adrenergic antagonist that is an effective antihypertensive. There is no role for **calcium sulphate**.

# Short Answer Questions
# Answers

## Question 7

A 70 year old male attends his GP's surgery. This patient is usually fit and well. He has mild osteoarthritis of his right knee, otherwise he has no medical history of note. Currently he complains of a recent onset severe headache and temporal pain, which is exacerbated when he combs his hair. He thinks that his vision has recently deteriorated on the same side that he experiences the pain and headaches.

A) What is the most likely diagnosis?
**Temporal arteritis**/Cranial arteritis/Giant cell arteritis (synonyms). (1 mark)

B) What blood test results should you see before initiating treatment? (1 mark)
**None**. This is a medical emergency that must be treated immediately based on the clinical scenario. An ESR/CRP ratio can be measured, but it can take days for the results to be returned. (1 mark)

C) What is the standard treatment for your diagnosis? (2 marks)
**High dose prednisolone** (glucocorticoid analogue). (2 marks)
It is not unusual to start with a 60mg prednisolone dose.

D) Name four side-effects of long term use of this class of drug. (4 marks)
The patient will develop **Cushingoid features**. The list of such features is extensive and includes:

Moon face
Buffalo hump
Abdominal striae
Poor wound healing
Predisposition to infection
Development of diabetes mellitus/hyperglycaemia

Polyuria and polydipsia
Muscle wasting
Central weight gain
Depression
Psychosis
Insomnia
Amenorrhea
Poor libido
Growth arrest
Back pain
Bruising
Hypertension

(1 mark for each correct answer – maximum of 4 marks).

E) Name three hyperglycaemic hormones. (3 marks)
Any three of:
**Cortisol**
**Growth hormone**
**Noradrenaline/adrenaline**
**Thyroid hormones**
**Glucagon**
**Lactogen**
**Oestrogens**
**Progestogens**

F) This unfortunate patient develops type 2 diabetes mellitus. He requires insulin for the management of his diabetes mellitus. Name one short acting and one long acting insulin or insulin analogue. (2 marks)
Common examples are:
**Actrapid** (Humulin S) is a short acting insulin.

**Novorapid** is a short acting insulin analogue.

**Human ultratard** is a long acting insulin.
**Glargine** is a long acting insulin analogue.

G) This patient continues to be unfortunate and develops an allergy to one of his insulin preparations. He suffers anaphylaxis in front of you. What drug would you use in his immediate management? What concentration/dilution would you use and how would you administer it? (3 marks)

**Adrenaline**
1/1000 dilution (1g adrenaline in 1000mls of fluid)
Intramuscularly

H) After taking a more careful history from this patient you note that he is atopic. He suffers from hayfever in the summer time and takes nasal Beconase and Benadryl. What are these drugs and what are their mechanisms of action? (4 marks)

Beconase (beclametasone) is a nasal steroid and so is a **corticosteroid analogue** (or **glucocorticoid analogue**) in the prednisolone class. It acts as a **local anti-inflammatory agent**.

Benadryl (diphenhydramine) is an **antihistamine**. It blocks the **H1 receptor**, so diminishing the effects of the released histamine.

# Viva Voce Questions:
# Answers

41) Aspirin, diclofenac and ibuprofen belong to which group of drugs?
NSAIDS (or analgesics).

42) For what is carbimazole used?
To treat hyperthyroidism; it inhibits thyroid peroxidase. Thyroid peroxidase is the key enzyme in thyroid hormone synthesis.

43) Which radioactive drug can be used to treat hyperthyroidism? What notorious risk is associated with this drug?
$^{131}$I. There is a risk of overtreatment that results in hypothyroidism.

44) Which receptors does labetalol block?
β adrenergic receptors and α adrenergic receptors.

45) Which has greater bioavailability a sublingual drug or an intravenous drug?
An intravenous drug because bioavailability is defined by the amount of the drug that enters the venous system. The intravenous drug has 100% bioavailability.

46) Is warfarin normally predominantly protein bound or non-protein bound in plasma?
99% protein bound.

47) What class of antibiotics are vancomycin and teicoplanin?
Glycopeptides.

48) Name three statins.
Popular answers include simvastatin, atorvastatin, pravastatin and rosuvastatin.

49) What type of drug is propranolol and what is its modern equivalent?
β adrenergic receptor antagonist. Commonly termed a *"beta blocker."*

50) In liver failure does the free circulating level of a patient's phenytoin rise or fall?

Rise. It is usually protein bound, however, during liver failure less protein is made and secreted into the circulation, so more of the phenytoin is unbound.

51) Do children metabolize hepatically degraded drugs faster or more slowly than adults? Why?

Faster. Because a greater proportion of a child's total body mass is liver. In addition, children have a proportionately higher hepatic blood flow.

52) Phenytoin, carbamazepine and sodium valproate are anticonvulsants. What common mechanism of action do they share?

They are sodium channel blockers. The voltage gated sodium channels in the cell membrane of nerve cells are inhibited, thus the development of action potentials is also inhibited.

53) How does imipramine work?

It is a tricyclic antidepressant. Its main action is to inhibit noradrenaline re-uptake in the CNS, raising noradrenaline levels at synapses.

54) Name three classes of drugs that you can use to treat depression.

Any three of:

Tricyclic antidepressants

Selective serotonin reuptake inhibitors.

Serotonin and noradrenaline reuptake inhibitors.

Monoamine oxidase inhibitors.

Atypical antidepressants.

# Appendix 1
## Glossary of Commonly Used Abbreviations in Prescriptions:

| Abbreviation | Latin | Meaning |
|---|---|---|
| **ad** | ad | up to |
| **a.c.** | ante cibum | before meals |
| **ad lib.** | ad libitum | use as much as is desired; freely |
| **admov.** | admove | apply |
| **agit** | agita | stir/shake |
| **alt. h.** | alternis horis | every other hour |
| **amp** | | ampule |
| **amt** | | amount |
| **aq** | aqua | water |
| **A.T.C.** | | around the clock |
| **bis** | bis | twice |
| **b.d./b.i.d.** | bis in die | twice daily |
| **B.M.** | | bowel movement |
| **BNF** | | British National Formulary |
| **bol.** | bolus | as a large single dose (usually intravenously) |
| **B.S.** | | blood sugar |
| **B.S.A** | | body surface areas |
| **BUCC** | bucca | inside cheek |
| **cap., caps.** | capsula | capsule |
| **c, c.** | cum | with |
| **cib.** | cibus | food |
| **cc** | cum cibo | with food, (and also *cubic centimetre*) |
| **cf** | | with food |
| **comp.** | | compound |

| cr., crm | | cream |
|---|---|---|
| CST | | Continue same treatment |
| D5W | | dextrose 5% solution (sometimes written as D₅W) |
| D5NS | | dextrose 5% in normal saline (0.9%) |
| D.A.W. | | dispense as written |
| dc, D/C, disc | | discontinue or discharge |
| dieb. alt. | diebus alternis | every other day |
| dil. | | dilute |
| disp. | | dispersible or dispense |
| div. | | divide |
| D.W. | | distilled water |
| elix. | | elixir |
| e.m.p. | ex modo prescripto | as directed |
| emuls. | emulsum | emulsion |
| et | et | and |
| eod | | every other day |
| ex aq | ex aqua | in water |
| fl., fld. | | fluid |
| ft. | fiat | make; let it be made |
| g | | gram |
| gtt(s) | gutta(e) | drop(s) |
| H | | hypodermic |
| h, hr | hora | hour |
| h.s. | hora somni | at bedtime |
| ID | | intradermal |
| IJ, inj | injectio | injection |
| IM | | intramuscular (injection) |

| IN | | intranasal |
|---|---|---|
| IP | | intraperitoneal |
| IU | | <u>international unit</u> |
| IV | | intravenous |
| lin | linimentum | liniment |
| liq | liquor | solution |
| lot. | | lotion |
| mane | mane | in the morning |
| m, min | minimum | a minimum |
| mcg | | microgram |
| mEq | | milliequivalent |
| mg | | milligram |
| mitte | mitte | send |
| mL | | millilitre |
| MS | | morphine sulphate or magnesium sulphate |
| MSO4 | | morphine sulphate |
| nebul | nebula | a spray |
| N.M.T. | | not more than |
| noct. | nocte | at night |
| non rep. | non repetatur | no repeats |
| NS | | normal saline (0.9%) |
| 1/2NS | | half normal saline (0.45%) |
| N.T.E. | | not to exceed |
| od | omne in die | every day/once daily |
| om | omne mane | every morning |
| on | omne nocte | every night |
| o.p.d. | | once per day |
| oz | | ounce |

| per | per | by or through |
|---|---|---|
| p.c. | post cibum | after meals |
| p.m. | post meridiem | evening or afternoon |
| p.o. | per os | by mouth or orally |
| p.r. | per rectum | by rectum |
| PRN, prn | pro re nata | as needed |
| PV | per vaginam | via the vagina |
| q | quaque | every |
| q.a.d. | quoque alternis die | every other day |
| q.a.m. | quaque die ante meridiem | every day before noon |
| q.d.s. | quater die sumendus | four times a day |
| q.p.m. | quaque die post meridiem | every day after noon |
| q.h. | quaque hora | every hour |
| q.h.s. | quaque hora somni | every night at bedtime |
| q.1h, q.1° | quaque 1 hora | every 1 hour; (can replace "1" with other numbers) |
| q.d., q1d | quaque die | every day |
| q.i.d. | quattuor in die | four times a day |
| q4PM | | at 4pm |
| q.o.d. | | every other day |
| qqh | quater quaque hora | every four hours |
| q.s. | quantum sufficiat | a sufficient quantity |

| QWK | | every week |
|---|---|---|
| R | | rectal |
| rep., rept. | repetatur | repeats |
| s | sine | without |
| SC, subc, subcut, subq, SQ | | subcutaneous |
| SL | | sublingually, under the tongue |
| sol | solutio | solution |
| ss | semis | one half or sliding scale |
| SSI, SSRI | | sliding scale insulin or sliding scale regular insulin |
| stat | statim | at once, immediately |
| supp | suppositorium | suppository |
| susp | | suspension |
| syr | syrupus | syrup |
| tab | tabella | tablet |
| t.d.s. | ter die sumendum | three times a day |
| t.i.d. | ter in die | three times a day |
| t.i.w. | | three times a week |
| top. | | topical |
| T.P.N., tpn | | total parenteral nutrition |
| tr, tinc., tinct. | | tincture |
| tsp | | teaspoon |
| u.d., ut. dict. | ut dictum | as directed |
| ung. | unguentum | ointment |
| vag | | vaginally |

| w | | with |
|---|---|---|
| wf | | with food (with meals) |
| w/o | | without |
| Y.O. | | years old |

# Appendix 2
# Commonly Prescribed Medicines:

| Drug | Typical Prescription |
|------|---------------------|
| Amlodipine | 10mg od po |
| Amoxycillin | 250mg tds po |
| Ampicillin | 500mg qds po |
| Aspirin | 300mg od po |
|  | 75mg od po |
| Atenolol | 50mg od po |
| Bendroflumethiazide | 2.5mg od po (mane) |
| Captopril | 12.5mg bd po |
| Cerazette | 1 tablet od po |
| Co-amoxiclav | For 250/125: 1 tablet tds po |
| Co-codamol | For 8/500: 1-2 tablets qds po |
| Codeine | 30mg po prn |
| Cyclizine | 50mg tds po |
| Dextrose (5%) | 1 litre/8 hours iv |
| Dextrose (5%) + 40mmol KCl | 1 litre/8 hours iv |
| Diamorphine | 5-10mg po prn |
| Diclofenac | 50mg tds po |
| Diltiazem | 60mg tds po |
| Enalapril | 20mg od po |
| Flucloxacillin | 500mg qds po |
| Fluoxetine | 20mg od po |
| Frusemide/furosemide | 40mg od po |
| Glyceryl trinitrate | 300 micrograms s/l prn |

| | |
|---|---|
| **Haloperidol** | 2-10mg im prn |
| **Hyoscine butylbromide** | 20mg qds po |
| **Ibuprofen** | 400mg qds po |
| | 600mg qds po |
| **Lanzoprazole** | 30mg od po |
| **Levothyroxine** | 50 micrograms od po (mane) |
| | 100 micrograms od po |
| **Metoclopramide** | 10mg tds po |
| **Microgynon 30** | 1 tablet od po |
| **Naproxen** | 250mg qds po |
| **Normal saline** | 1 litre/8 hours iv |
| **Omeprazole** | 20mg od po |
| **Paracetamol** | 1g qds po |
| **Pravastatin** | 40mg od po (nocte) |
| **Prednisolone** | 20mg od po |
| | 10mg od po |
| **Prochlorperazine** | 10mg tds po |
| **Ramipril** | 1.25mg od po |
| | 5mg od po |
| **Salbutamol** | 100-200 micrograms (1-2 puffs) |
| **Simvastatin** | 40mg od po (nocte) |
| **Trimethoprim** | 200mg bd po |
| **Verapamil** | 80mg tds po |

www.ingramcontent.com/pod-product-compliance
Lightning Source LLC
Chambersburg PA
CBHW060550210326
41519CB00014B/3426

9 7 8 0 9 5 6 6 4 4 3 4 3